AF483749

FORMIDABLE PROBLEMS IN THERMAL PHYSICS

(Fully Solved)

Dr. Sunirmit Verma

ISBN
Paperback 979-8-89446-175-5
Hardcase 979-8-89446-277-6

What this Book is About

This book is a collection of convoluted problems created by me in the field of thermal physics. By convoluted, I mean they are modifications of problems found in standard physics textbooks on the topic which require more rigorous mathematical treatment. There are a total of 30 problems, and all of them involve mathematical expression generation, that is, some parameters are given in question in symbolic form instead of numerical values, and what is required to be found out is to be done in the form of a mathematical expression in terms of mentioned parameters. A general procedure is followed while dealing with any problem on the subject, which is described ahead. All the problems are fully solved with every step of the solution explained meticulously. The aim of the book is to help readers develop mathematical skills to predict temperature fields in any system experiencing some sort of energy interaction (with other systems).

The general procedure for solving any problem in thermal physics is:

STEP 1:   Write down the law of energy conservation for a general, differential volume of each object involved. This law in this context is: Rate of energy entering minus rate of energy leaving equals the rate of change of internal energy of the differential element.

STEP 2:   Step 1 shall yield a set of differential equations. Identify the relevant boundary and/or initial conditions

STEP 3:   Solve the set of differential equations

STEP 4:   Step 3 shall yield the temperature field in each object. Write down the rate of entropy generation for a general, differential volume of each object involved. It is rate of energy entering, minus rate of energy leaving divided by absolute instantaneous temperature of the differential element. Integrate this with respect to space for each object, sum all the entropy generation rates to get the net rate. Integrate this net rate with respect to time to obtain entropy generated in a given time interval

Acknowledgement

I dedicate this book to my parents, who consecrated their lives to the sole aim of filling my stay in this Earthly realm with exuberance and shielding me from all sorrows out there.

Question 1

A solid body of mass m and specific heat capacity c_p loses heat via radiation alone to its surroundings which are at an initial temperature T_a. Initial temperature of body is T_i, and surroundings are very large. Find body temperature as a function of time. The surface area and emissivity of the body are A and ε respectively. Emissivity is temperature dependent as per the relation: $\varepsilon = \dfrac{\alpha}{T}$ where α is a material specific constant and T is instantaneous temperature. Neglect temperature gradients inside the body. The Stefan Boltzmann constant is σ.

Solution

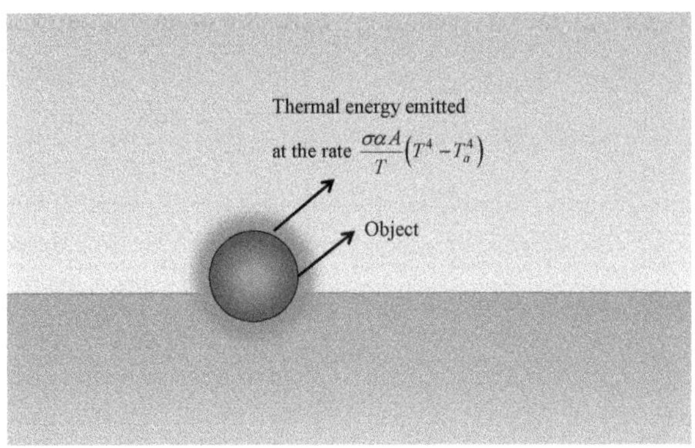

Applying law of energy conservation for the body at a general time t

$$\frac{d}{dt}\left(mc_pT\right) = -\sigma\varepsilon A\left(T^4 - T_a^4\right)$$

$$\Rightarrow mc_p\frac{dT}{dt} = -\frac{\sigma\alpha A}{T}\left(T^4 - T_a^4\right) \tag{1}$$

Separating the variables and integrating both sides

$$\int_{T_i}^{T}\frac{T\cdot dT}{T^4 - T_a^4} = -\int_{0}^{t}\frac{\sigma\alpha A}{mc_p}\,dt \tag{2}$$

Since surroundings are large, their temperature T_a can be assumed to remain constant. The integral appearing on the left-hand side of above equation can be simplified as shown below

$$\text{Integral} = \int_{T_i}^{T}\frac{T\cdot dT}{\left(T^2 - T_a^2\right)\left(T^2 + T_a^2\right)} \tag{3}$$

Put $T^2 = v \Rightarrow 2T\cdot dT = dv$

$$\text{Integral} = \int_{T_i^2}^{T^2}\frac{dv}{2\left(v - T_a^2\right)\left(v + T_a^2\right)} \tag{4}$$

Decomposing the integrand into partial fractions

$$\text{Integral} = \int_{T_i^2}^{T^2}\left[\frac{1}{v - T_a^2} - \frac{1}{v + T_a^2}\right]\frac{dv}{4T_a^2} \tag{5}$$

Simplifying

$$\text{Integral} = \frac{1}{4T_a^2}\left[\int\limits_{T_i^2}^{T^2}\frac{dv}{v-T_a^2} - \int\limits_{T_i^2}^{T^2}\frac{dv}{v+T_a^2}\right]$$

$$\Rightarrow \text{Integral} = \frac{1}{4T_a^2}\left[\ln\left|\frac{T^2-T_a^2}{T_i^2-T_a^2}\right| - \ln\left|\frac{T^2+T_a^2}{T_i^2+T_a^2}\right|\right] \tag{6}$$

Feeding equation 6 in 2

$$\frac{1}{4T_a^2}\left[\ln\left|\frac{T^2-T_a^2}{T_i^2-T_a^2}\right| - \ln\left|\frac{T^2+T_a^2}{T_i^2+T_a^2}\right|\right] = -\frac{\sigma\alpha At}{mc_p}$$

$$\Rightarrow \ln\left|\frac{T^2-T_a^2}{T_i^2-T_a^2}\right| - \ln\left|\frac{T^2+T_a^2}{T_i^2+T_a^2}\right| = -\frac{4T_a^2\sigma\alpha At}{mc_p}$$

$$\Rightarrow \ln\left|\frac{T^2-T_a^2}{T_i^2-T_a^2}\cdot\frac{T_i^2+T_a^2}{T^2+T_a^2}\right| = -\frac{4T_a^2\sigma\alpha At}{mc_p} \tag{7}$$

Simplifying

$$\frac{T^2-T_a^2}{T^2+T_a^2} = \left(\frac{T_i^2-T_a^2}{T_i^2+T_a^2}\right)e^{-\frac{4T_a^2\sigma\alpha At}{mc_p}} \tag{8}$$

$$T^2-T_a^2 = \left(\frac{T_i^2-T_a^2}{T_i^2+T_a^2}\right)\left(T^2+T_a^2\right)e^{-\frac{4T_a^2\sigma\alpha At}{mc_p}}$$

$$\Rightarrow T^2-T_a^2 = \left(\frac{T_i^2-T_a^2}{T_i^2+T_a^2}\right)T^2 e^{-\frac{4T_a^2\sigma\alpha At}{mc_p}} + \left(\frac{T_i^2-T_a^2}{T_i^2+T_a^2}\right)T_a^2 e^{-\frac{4T_a^2\sigma\alpha At}{mc_p}} \tag{9}$$

$$T^2\left[1-\left\{\left(\frac{T_i^2-T_a^2}{T_i^2+T_a^2}\right)e^{-\frac{4T_a^2\sigma\alpha At}{mc_p}}\right\}\right]=T_a^2\left[1+\left\{\left(\frac{T_i^2-T_a^2}{T_i^2+T_a^2}\right)e^{-\frac{4T_a^2\sigma\alpha At}{mc_p}}\right\}\right]$$

$$\Rightarrow T=T_a\sqrt{\frac{\left\{T_i^2+T_a^2\right\}+\left\{\left(T_i^2-T_a^2\right)e^{-\frac{4T_a^2\sigma\alpha At}{mc_p}}\right\}}{\left\{T_i^2+T_a^2\right\}-\left\{\left(T_i^2-T_a^2\right)e^{-\frac{4T_a^2\sigma\alpha At}{mc_p}}\right\}}}$$

$$(10)$$

Question 2

The Newton's law of cooling states that for small temperature differences with the surroundings, rate of cooling of a solid body is proportional to the first power of its temperature difference with the surroundings. It can be derived from the Stefan Boltzmann law, by neglecting 2nd and higher power terms of this difference. Considering radiation only corresponding to the most dominant wavelength and neglecting convection, derive an alternative form of Newton's law of cooling for a blackbody of mass m, specific heat capacity c_p and surface area A. The Planck's constant is h, Boltzmann constant is k, speed of light in vacuum is c, Wein's displacement constant is b. The surrounding temperature (assumed constant in this derivation because of large size of surroundings) is T_a. Neglect temperature gradients inside the body.

Solution

The Planck's law of radiation gives the radiant energy emission rate per unit area for a black body at temperature T, corresponding to a wavelength λ and is mentioned below

$$\dot{E}_{\lambda b} = \frac{2\pi c^2 h \lambda^{-5}}{e^{\frac{ch}{k\lambda T}} - 1}$$

(1)

The wavelength at which this quantity is maximum is known as the most dominant wavelength for the blackbody at that temperature and is given by the following equation known as the Wein's displacement law

$$\lambda_d = \frac{b}{T}$$

(2)

Feeding this critical value of λ in equation 1 gives the corresponding maximum radiant energy emission rate

$$\left(\dot{E}_{\lambda b}\right)_{\text{max}} = \frac{2\pi c^2 h b^{-5} T^5}{e^{\frac{ch}{kb}} - 1}$$

(3)

Since we are to consider emission only corresponding to this wavelength, therefore, in this question, radiant energy emission rate is to be taken proportional to 5th power of absolute body temperature. Applying law of energy conservation for the body at a general time t

$$\frac{d\left(mc_p T\right)}{dt} = -\frac{2\pi c^2 h A b^{-5}\left(T^5 - T_a^5\right)}{e^{\frac{ch}{kb}} - 1}$$

$$\Rightarrow \frac{dT}{dt} = -\frac{2\pi c^2 h A b^{-5}\left(T^5 - T_a^5\right)}{mc_p\left(e^{\frac{ch}{kb}} - 1\right)}$$

(4)

The term $\left(T^5 - T_a^5\right)$ can be rewritten under the condition of $T\text{-}T_a$ being very small. Let $T\text{-}T_a$ be x such that x^2, x^3, x^4 and other higher power terms of x can be ignored. The following algebraic factorisation can be used

$$X^n - Y^n = (X - Y)(X^{n-1} + X^{n-2}Y + X^{n-3}Y^2 + \dots\dots\dots XY^{n-2} + Y^{n-1})$$

where $X, Y \in R$, $n \in N$ $\qquad(5)$

According to equation 5, the term $\left(T^5 - T_a^5\right)$ can be expressed as

$$T^5 - T_a^5 = (T - T_a)(T^4 + T^3 T_a + T^2 T_a^2 + T T_a^3 + T_a^4) \qquad(6)$$

$$T^5 - T_a^5 = (T - T_a)(T^4 + T_a^4 - 2T^2 T_a^2 + T^3 T_a + 3T^2 T_a^2 + T T_a^3) \qquad(7)$$

$$T^5 - T_a^5 = (T - T_a)\left[\left(T^2 - T_a^2\right)^2 + T T_a\left(T^2 + 3T T_a + T_a^2\right)\right] \qquad(8)$$

$$T^5 - T_a^5 = x\left[\begin{array}{l}\left\{x^2 + T_a^2 + 2xT_a - T_a^2\right\}^2 + \\ \left\{(x + T_a)T_a\left\{x^2 + T_a^2 + 2xT_a + 3(x + T_a)T_a + T_a^2\right\}\right\}\end{array}\right] \qquad(9)$$

Ignoring x^2 and higher power terms of x

$$T^5 - T_a^5 = x\left[\left\{2xT_a\right\}^2 + \left\{(x + T_a)T_a\left(T_a^2 + 2xT_a + 3T_a x + 3T_a^2 + T_a^2\right)\right\}\right] \qquad(10)$$

Further, since x is very small, $T_a + x \approx T_a$

$$T^5 - T_a^5 = x\left[\left\{4x^2 T_a^2\right\} + \left\{T_a^2\left(5T_a^2 + 5xT_a\right)\right\}\right] \qquad(11)$$

$$T^5 - T_a^5 = x\left[5T_a^3\left(x + T_a\right)\right] \qquad(12)$$

$$T^5 - T_a^5 = 5T_a^4 x \qquad(13)$$

Feeding equation 13 in 4

$$\frac{dT}{dt} = -\frac{10T_a^4 \pi c^2 hAb^{-5}\left(T-T_a\right)}{mc_p\left(e^{\frac{ch}{kb}}-1\right)}$$

(14)

Hence, it is proved that

$$\frac{dT}{dt} \propto \left(T-T_a\right)$$

where the proportionality constant is $\dfrac{-10T_a^4 \pi c^2 hAb^{-5}}{mc_p\left(e^{\frac{ch}{kb}}-1\right)}$

(15)

Question 3

A solid body of mass m and specific heat capacity c_p loses heat via convection alone to its surroundings which are at an initial temperature T_a. Initial temperature of body is T_i, and surroundings are very large. Find body temperature as a function of time. The surface area of the body is A. Convection coefficient with the surroundings h is temperature dependent: $h = \mu T$: where μ is a material specific constant and T is instantaneous temperature. Neglect temperature gradients inside the body.

Solution

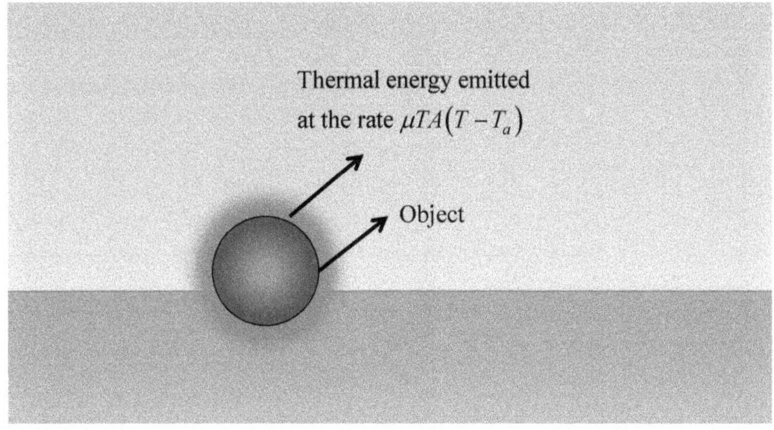

Applying law of energy conservation for the body at a general time t

$$\frac{d\left(mc_p T\right)}{dt} = -hA\left(T - T_a\right)$$

$$\Rightarrow mc_p \frac{dT}{dt} = -\mu TA\left(T - T_a\right) \tag{1}$$

Since surroundings are large, their temperature T_a can be assumed to remain constant. Equation 1 can be written in condensed form as

$$\frac{dT}{dt} = \alpha T - \beta T^2 \tag{2}$$

where $\alpha = \dfrac{\mu T_a A}{mc_p}, \beta = \dfrac{\mu A}{mc_p}$. Separating the variables and integrating both sides

$$\int_{T_i}^{T} \frac{dT}{\alpha T - \beta T^2} = \int_0^t dt \tag{3}$$

Simplifying

$$\int_{T_i}^{T} \frac{dT}{\beta T^2 - \alpha T} = -\int_0^t dt$$

$$\Rightarrow \int_{T_i}^{T} \frac{dT}{\beta\left(T^2 - \dfrac{\alpha}{\beta}T\right)} = -\int_0^t dt \tag{4}$$

$$\int_{T_i}^{T} \frac{dT}{T^2 - \dfrac{\alpha}{\beta}T + \dfrac{\alpha^2}{4\beta^2} - \dfrac{\alpha^2}{4\beta^2}} = -\beta t \tag{5}$$

$$\int_{T_i}^{T} \frac{dT}{\left(T - \dfrac{\alpha}{2\beta}\right)^2 - \left(\dfrac{\alpha}{2\beta}\right)^2} = -\beta t \tag{6}$$

Using the standard result $\displaystyle\int \frac{dZ}{Z^2 - S^2} = \frac{1}{2S} \ln\left|\frac{Z-S}{Z+S}\right|$

$$\frac{1}{\dfrac{\alpha}{\beta}} \ln\left|\frac{T - \dfrac{\alpha}{\beta}}{T}\right| = -\beta t$$

$$\Rightarrow \ln\left|\frac{T - \dfrac{\alpha}{\beta}}{T}\right| = -\alpha t \tag{7}$$

$$1 - \frac{\alpha}{T\beta} = e^{-\alpha t}$$

$$\Rightarrow \frac{\alpha}{T\beta} = 1 - e^{-\alpha t}$$

$$\Rightarrow T = \frac{\alpha}{\beta\left(1 - e^{-\alpha t}\right)} \tag{8}$$

Expanding back α, β

$$T = \frac{T_a}{1 - e^{-\frac{\mu T_a A t}{m c_p}}} \tag{9}$$

Question 4

❖❖❖

A solid body of mass m and surface area A loses heat via convection alone to its surroundings which are at an initial temperature T_a. Initial temperature of body is Ti, and surroundings are very large. The convective heat transfer coefficient between body and surroundings is h. The specific heat capacity of the body's material is a linear function of temperature $[c_p = a_1 + b_1 T]$ where a_1, b_1 are material specific constants.Find the body temperature as a function of time. Neglect temperature gradients inside the body.

Solution

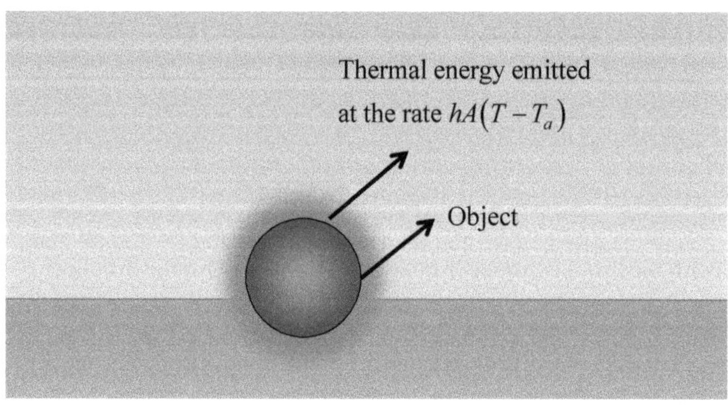

Applying law of energy conservation for the body at a general time t

$$\frac{d(mc\;T)}{dt} = -hA(T-T)$$

(1)

$$m\frac{d(c_p T)}{dt} = -hA(T-T_a)$$

(2)

Feeding the mentioned expression ofspecific heat capacity in equation 2

$$\frac{d}{dt}\{(a_1 + b_1 T)T\} = -\frac{hA}{m}(T-T_a)$$

$$\Rightarrow (a_1 + b_1 T)\frac{dT}{dt} + b_1 T\frac{dT}{dt} = -\frac{hA}{m}(T-T_a)$$

$$\Rightarrow (a_1 + 2b_1 T)\frac{dT}{dt} = -\frac{hA}{m}(T-T_a)$$

(3)

Separating the variables and integrating both sides

$$\int_{T_i}^{T}\frac{(a_1 + 2b_1 T)\cdot dT}{T-T_a} = -\int_{0}^{t}\frac{hA\cdot dt}{m}$$

(4)

Simplifying

$$a_1\int_{T_i}^{T}\frac{dT}{T-T_a} + 2b_1\int_{T_i}^{T}\frac{T\cdot dT}{T-T_a} = -\int_{0}^{t}\frac{hA\cdot dt}{m}$$

(5)

$$a_1\int_{T_i}^{T}\frac{dT}{T-T_a} + 2b_1\int_{T_i}^{T}\frac{(T-T_a+T_a)\cdot dT}{T-T_a} = -\int_{0}^{t}\frac{hA\cdot dt}{m}$$

(6)

$$a_1 \int_{T_i}^{T} \frac{dT}{T - T_a} + 2b_1 \int_{T_i}^{T} dT + 2b_1 T_a \int_{T_i}^{T} \frac{dT}{T - T_a} = - \int_{0}^{t} \frac{hA \cdot dt}{m}$$

$$\Rightarrow \left(a_1 + 2b_1 T_a \right) \int_{T_i}^{T} \frac{dT}{T - T_a} + 2b_1 \int_{T_i}^{T} dT = - \int_{0}^{t} \frac{hA \cdot dt}{m} \tag{7}$$

$$\left(a_1 + 2b_1 T_a \right) \ln \left| \frac{T - T_a}{T_i - T_a} \right| + 2b_1 \left(T - T_i \right) = - \frac{hAt}{m} \tag{8}$$

Question 5

A solid body loses heat via convection alone to its surroundings which are at an initial temperature T_a. Initial temperature of body is T_i, and surroundings are very large. As the body loses heat, it also undergoes sublimation and turns to vapour as per the law: Instantaneous mass loss rate $= -h_{sub}A(T-T_a)$. Initial mass of the body is m_i. The surface area and specific heat capacity of the body are A and c_p respectively. Its convection coefficient with the surroundings is h. Find body mass as a function of time. Neglect temperature gradients inside the body.

Solution

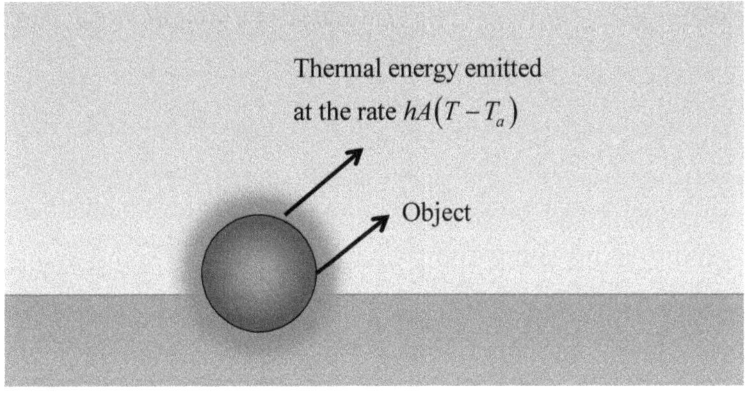

Applying law of energy conservation for the body at a general time t

$$\frac{d}{dt}\left(mc_pT\right) = -hA\left(T - T_a\right)$$

$$\Rightarrow c_p\frac{d\left(mT\right)}{dt} = -hA\left(T - T_a\right) \tag{1}$$

As per the given information,

$$\frac{dm}{dt} = -h_{sub}A\left(T - T_a\right) \tag{2}$$

Equations 1 and 2 are coupled ordinary differential equations. Feeding T from equation 2 in 1

$$\frac{d}{dt}\left[m\left(T_a - \frac{\dfrac{dm}{dt}}{h_{sub}A}\right)\right] = \frac{h}{h_{sub}c_p}\frac{dm}{dt}$$

$$\Rightarrow d\left[m\left(T_a - \frac{\dfrac{dm}{dt}}{h_{sub}A}\right)\right] = \frac{h}{h_{sub}c_p}dm \tag{3}$$

Integrating both sides

$$\int d\left[m\left(T_a - \frac{\dfrac{dm}{dt}}{h_{sub}A}\right)\right] = \int \frac{h}{h_{sub}c_p}dm$$

$$\Rightarrow m\left(T_a - \frac{\dfrac{dm}{dt}}{h_{sub}A}\right) = \frac{hm}{h_{sub}c_p} + C_0 \tag{4}$$

Where C_0 is an arbitrary integration constant. Equation 4 can be rearranged as

$$\frac{dm}{dt} = \alpha - \frac{\beta}{m} \tag{5}$$

Where $\alpha = h_{sub} A T_a - \dfrac{hA}{c_p}, \beta = C_0 h_{sub} A$

Separating the variables and integrating both sides

$$\int_{m_i}^{m} \frac{m \cdot dm}{\alpha m - \beta} = \int_0^t dt \tag{6}$$

Simplifying

$$\frac{1}{\alpha} \int_{m_i}^{m} \frac{\alpha m \cdot dm}{\alpha m - \beta} = t$$

$$\Rightarrow \frac{1}{\alpha} \int_{m_i}^{m} \frac{(\alpha m - \beta + \beta) \cdot dm}{\alpha m - \beta} = t$$

$$\Rightarrow \int_{m_i}^{m} dm + \beta \int_{m_i}^{m} \frac{dm}{\alpha m - \beta} = \alpha t$$

$$\Rightarrow m - m_i + \frac{\beta}{\alpha} \ln \left| \frac{\alpha m - \beta}{\alpha m_i - \beta} \right| = \alpha t \tag{7}$$

Expanding back α, β

$$m - m_i + \left(\frac{C_0 h_{sub}}{h_{sub} T_a - \dfrac{h}{c_p}} \right) \ln \left| \frac{\left(h_{sub} A T_a - \dfrac{hA}{c_p} \right) m - C_0 h_{sub} A}{\left(h_{sub} A T_a - \dfrac{hA}{c_p} \right) m_i - C_0 h_{sub} A} \right|$$

$$= \left(h_{sub} A T_a - \frac{hA}{c_p} \right) t \tag{8}$$

The expression obtained by far contains an unknown integration constant C_0. It needs to be found out by the initial condition $T_{t=0} = T_i$. For that, an expression for T is needed. This can be done using equation 2 and 5

$$\alpha - \frac{\beta}{m} = -h_{sub} A (T - T_a)$$

$$\Rightarrow h_{sub} A T_a - \frac{hA}{c_p} - \frac{C_0 h_{sub} A}{m} = -h_{sub} A (T - T_a)$$

$$\Rightarrow T = \frac{h}{h_{sub} c_p} + \frac{C_0}{m}$$

$$(9)$$

Although an explicit function of T in terms of t is not obtainable, still equation 9 can be used to find C_0. The fact that at $t=0$. $T=T_i$ and $m=mi$ helps

$$T_i = \frac{h}{h_{sub} c_p} + \frac{C_0}{m_i}$$

$$\Rightarrow C_0 = m_i T_i - \frac{hm_i}{h_{sub} c_p}$$

$$(10)$$

Feeding equation 10 in 8 completes the expression that was sought

$$m - m_i + \left(\frac{m_i T_i h_{sub} - \dfrac{hm_i}{c_p}}{h_{sub} T_a - \dfrac{h}{c_p}} \right) \ln \left| \frac{h_{sub} A T_a m - \dfrac{hAm}{c_p} - m_i T_i h_{sub} A + \dfrac{hm_i A}{c_p}}{h_{sub} A m_i (T_a - T_i)} \right|$$

$$= \left(h_{sub} A T_a - \frac{hA}{c_p} \right) t$$

$$(11)$$

Question 6

❖ ❖ ❖

2 solid bodiexs of masses m_1, m_2 and specific heat capacities c_{p1}, c_{p2} exchange heat as per the law $\dot{Q} = a\Delta T$ where ΔT is the instantaneous temperature difference between them. At *t=0*, their temperatures are T_{1i} and T_{2i}. Find net entropy generated in the heat transfer process. Neglect temperature gradients inside each body.

Solution

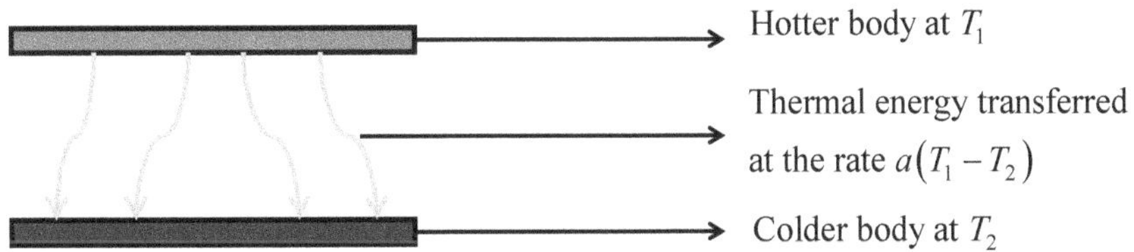

Applying law of energy conservation for the two bodies at a general time *t,*

$$\frac{d\left(m_1 c_{p1} T_1\right)}{dt} = -a\left(T_1 - T_2\right)$$

$$\Rightarrow \frac{dT_1}{dt} = -\frac{a}{m_1 c_{p1}}\left(T_1 - T_2\right) \tag{1}$$

$$\frac{d\left(m_2 c_{p2} T_2\right)}{dt} = a\left(T_1 - T_2\right)$$

$$\Rightarrow \frac{dT_2}{dt} = \frac{a}{m_2 c_{p2}}\left(T_1 - T_2\right) \tag{2}$$

Equations 1 and 2 are intercoupled ordinary differential equations. One of the two dependent variables (T_1 and T_2) needs to be eliminated to proceed further

Substituting T_1 from Equation 2 in 1

$$\frac{d}{dt}\left[T_2 + \frac{m_2 c_{p2}}{a}\frac{dT_2}{dt}\right] = -\frac{m_2 c_{p2}}{m_1 c_{p1}}\frac{dT_2}{dt} \tag{3}$$

Simplifying

$$\frac{d^2 T_2}{dt^2} + \left[\frac{a\left(m_1 c_{p1} + m_2 c_{p2}\right)}{m_1 c_{p1} m_2 c_{p2}}\right]\frac{dT_2}{dt} = 0 \tag{4}$$

Separating the variables and integrating both sides

$$\frac{dT_2}{dt} + \chi T_2 = C_1 \tag{5}$$

Where $\lambda = \dfrac{a\left(m_1 c_{p1} + m_2 c_{p2}\right)}{m_1 c_{p1} m_2 c_{p2}}$ and C_1 is an arbitrary integration constant. Separating the variables and integrating both sides

$$\int_{T_{2i}}^{T_2} \frac{dT_2}{C_1 - \chi T_2} = \int_0^t dt$$

$$\Rightarrow \frac{1}{-\chi} \ln\left(\frac{C_1 - \chi T_2}{C_1 - \chi T_{2i}}\right) = t$$

$$\Rightarrow \frac{C_1 - \chi T_2}{C_1 - \chi T_{2i}} = e^{-\chi t}$$

$$\Rightarrow T_2 = \frac{C_1}{\chi} - \left(\frac{C_1}{\chi} - T_{2i}\right) e^{-\chi t} \tag{6}$$

Feeding Equation 6 in 2

$$T_1 = \left\{\frac{C_1}{\chi}\right\} - \left\{\left(\frac{C_1}{\chi} - T_{2i}\right)e^{-\chi t}\right\} + \left\{\frac{m_2 c_{p2} \chi}{a}\left(\frac{C_1}{\chi} - T_{2i}\right)e^{-\chi t}\right\}$$

$$\Rightarrow T_1 = \left\{\frac{C_1}{\chi}\right\} + \left\{\left(\frac{C_1}{\chi} - T_{2i}\right)\left(\frac{\chi m_2 c_{p2}}{a} - 1\right)e^{-\chi t}\right\} \tag{7}$$

The constant C_1 has to be found by the initial condition $T_1(t{=}0){=}T_{1i}$

$$T_{1i} = \left\{\frac{C_1}{\chi}\right\} + \left\{\left(\frac{C_1}{\chi} - T_{2i}\right)\left(\frac{\chi m_2 c_{p2}}{a} - 1\right)\right\}$$

$$\Rightarrow T_{1i} = \left\{\frac{C_1}{\chi}\right\} + \left\{\left(\frac{C_1}{\chi}\right)\left(\frac{\chi m_2 c_{p2}}{a} - 1\right)\right\} - \left\{T_{2i}\left(\frac{\chi m_2 c_{p2}}{a} - 1\right)\right\}$$

$$\Rightarrow C_1 = \frac{a}{m_2 c_{p2}}\left[\{T_{1i}\} + \left\{T_{2i}\left(\frac{\chi m_2 c_{p2}}{a} - 1\right)\right\}\right] \tag{8}$$

Entropy generation rate in the differential time intrreval dt

$$d\dot{S} = -\frac{a(T_1 - T_2)}{T_1} + \frac{a(T_1 - T_2)}{T_2} = a(T_1 - T_2)\left(\frac{-1}{T_1} + \frac{1}{T_2}\right) = \frac{a(T_1 - T_2)^2}{T_1 T_2} \tag{9}$$

Using Equations 6 and 7 in 9

$$d\dot{S} = \frac{a\left[\left\{\frac{C_1}{\chi}\right\}+\left\{\left(\frac{C_1}{\chi}-T_{2i}\right)\left(\frac{\chi m_2 c_{p2}}{a}-1\right)e^{-\chi t}\right\}-\left\{\frac{C_1}{\chi}\right\}+\left\{\left(\frac{C_1}{\chi}-T_{2i}\right)e^{-\chi t}\right\}\right]^2}{\left[\left\{\frac{C_1}{\chi}\right\}+\left\{\left(\frac{C_1}{\chi}-T_{2i}\right)\left(\frac{\chi m_2 c_{p2}}{a}-1\right)e^{-\chi t}\right\}\right]\left[\left\{\frac{C_1}{\chi}\right\}-\left\{\left(\frac{C_1}{\chi}-T_{2i}\right)e^{-\chi t}\right\}\right]}$$

(10)

This can be condensed as

$$d\dot{S} = \frac{a\left(\frac{C_1}{\chi}-T_{2i}\right)^2\left(\frac{\chi m_2 c_{p2}}{a}\right)^2 e^{-2\chi t}}{\left[\left\{\frac{C_1}{\chi}\right\}+\left\{\left(\frac{C_1}{\chi}-T_{2i}\right)\left(\frac{\chi m_2 c_{p2}}{a}-1\right)e^{-\chi t}\right\}\right]\left[\left\{\frac{C_1}{\chi}\right\}-\left\{\left(\frac{C_1}{\chi}-T_{2i}\right)e^{-\chi t}\right\}\right]}$$

$$\Rightarrow d\dot{S} = \frac{a\left(\frac{C_1}{\chi}-T_{2i}\right)^2\left(\frac{\chi m_2 c_{p2}}{a}\right)^2 e^{-2\chi t}}{\left[\frac{C_1}{\chi}+pe^{-\chi t}\right]\left[\frac{C_1}{\chi}-qe^{-\chi t}\right]}$$

(11)

where p$=\left(\frac{C_1}{\chi}-T_{2i}\right)\left(\frac{\chi m_2 c_{p2}}{a}-1\right)$, q$=\frac{C_1}{\chi}-T_{2i}$.

To find the net entropy generated, the above expression needs to integrated with respect to time from t=0 upto the time when heat transfer stops, which is when T_1-T_2=0. From equations 6 and 7, an expression for T_1-T_2 can be obtained and it can be seen that this quantity attains 0 at $t \to \infty$. Thus,

$$S = \int dS = \int \frac{dS}{dt}dt = \int_0^\infty \frac{a\left(\frac{C_1}{\chi}-T_{2i}\right)^2\left(\frac{\chi m_2 c_{p2}}{a}\right)^2 e^{-2\chi t}}{\left[\frac{C_1}{\chi}+pe^{-\chi t}\right]\left[\frac{C_1}{\chi}-qe^{-\chi t}\right]}\cdot dt$$

(12)

Put $e^{-\chi t} = z, \Rightarrow -\chi e^{-\chi t}\, dt = dz$

$$S = a\left(\frac{C_1}{\chi} - T_{2i}\right)^2 \left(\frac{\chi m_2 c_{p2}}{a}\right)^2 \int_1^0 \frac{z^2 \cdot dz}{-\lambda z \left(\frac{C_1}{\chi} + pz\right)\left(\frac{C_1}{\chi} - qz\right)}$$

$$\Rightarrow S = -\frac{a}{\lambda}\left(\frac{C_1}{\chi} - T_{2i}\right)^2 \left(\frac{\chi m_2 c_{p2}}{a}\right)^2 \int_1^0 \frac{z \cdot dz}{\left(\frac{C_1}{\chi} + pz\right)\left(\frac{C_1}{\chi} - qz\right)} \tag{13}$$

Simplifying

$$S = \frac{-a}{\lambda}\left(\frac{C_1}{\chi} - T_{2i}\right)^2 \left(\frac{\chi m_2 c_{p2}}{a}\right)^2 \frac{1}{(p+q)} \int_1^0 \frac{(p+q)z \cdot dz}{\left(\frac{C_1}{\chi} + pz\right)\left(\frac{C_1}{\chi} - qz\right)}$$

$$= \frac{-a}{\lambda}\left(\frac{C_1}{\chi} - T_{2i}\right)^2 \left(\frac{\chi m_2 c_{p2}}{a}\right)^2 \frac{1}{(p+q)} \int_1^0 dz \left[\frac{1}{\frac{C_1}{\chi} - qz} - \frac{1}{\frac{C_1}{\chi} + pz}\right]$$

$$= \frac{-a}{\lambda}\left(\frac{C_1}{\chi} - T_{2i}\right)^2 \left(\frac{\chi m_2 c_{p2}}{a}\right)^2 \frac{1}{(p+q)} \left[-\frac{1}{q}\ln\left|\frac{C_1}{\chi} - qz\right| - \frac{1}{p}\ln\left|\frac{C_1}{\chi} + pz\right|\right]_1^0$$

$$= \frac{a}{\lambda}\left(\frac{C_1}{\chi} - T_{2i}\right)^2 \left(\frac{\chi m_2 c_{p2}}{a}\right)^2 \frac{1}{(p+q)} \left[\frac{1}{q}\ln\left|\frac{C_1}{\chi} - qz\right| + \frac{1}{p}\ln\left|\frac{C_1}{\chi} + pz\right|\right]_1^0 \tag{14}$$

$$S = \frac{a}{\lambda}\left(\frac{C_1}{\chi} - T_{2i}\right)^2 \left(\frac{\chi m_2 c_{p2}}{a}\right)^2 \frac{1}{(p+q)} \left[\frac{1}{q}\ln\left|\frac{C_1}{C_1 - \lambda q}\right| + \frac{1}{p}\ln\left|\frac{C_1}{C_1 + \lambda p}\right|\right] \tag{15}$$

Expanding back *p, q,*

$$S = m_2 c_{p2} \left(\frac{C_1}{\chi} - T_{2i} \right) \left[\begin{array}{c} \dfrac{\lambda}{C_1 - \lambda T_{2i}} \ln \left| \dfrac{C_1}{\lambda T_{2i}} \right| + \\[2ex] \dfrac{a\lambda}{\left(C_1 - \lambda T_{2i} \right)\left(\chi m_2 c_{p2} - a \right)} \ln \\[2ex] \left| \dfrac{a C_1}{\left\{ a C_1 \right\} + \left\{ \left(C_1 - \lambda T_{2i} \right)\left(\chi m_2 c_{p2} - a \right) \right\}} \right| \end{array} \right]$$

(16)

Feeding C_1 from equation 8 in equation 16

$$S = a \left(\frac{T_{1i} - T_{2i}}{\lambda} \right) \left[\begin{array}{c} \dfrac{\lambda m_2 c_{p2}}{a \left(T_{1i} - T_{2i} \right)} \ln \left| \dfrac{a T_{1i} + T_{2i} \chi m_2 c_{p2} - a T_{2i}}{\lambda T_{2i} m_2 c_{p2}} \right| + \\[2ex] \dfrac{\lambda m_2 c_{p2}}{\left(T_{1i} - T_{2i} \right)\left(\chi m_2 c_{p2} - a \right)} \ln \\[2ex] \left| \dfrac{a T_{1i} + T_{2i} \chi m_2 c_{p2} - a T_{2i}}{\left\{ a T_{1i} + T_{2i} \chi m_2 c_{p2} - a T_{2i} \right\} + \left\{ \left(T_{1i} - T_{2i} \right)\left(\chi m_2 c_{p2} - a \right) \right\}} \right| \end{array} \right]$$

(17)

Expanding back λ

$$S = \frac{m_1 c_{p1} m_2 c_{p2} \left(T_{1i} - T_{2i} \right)}{m_1 c_{p1} + m_2 c_{p2}} \left[\begin{array}{c} \dfrac{m_1 c_{p1} + m_2 c_{p2}}{m_1 c_{p1} \left(T_{1i} - T_{2i} \right)} \ln \left| \dfrac{m_1 c_{p1} T_{1i} + m_2 c_{p2} T_{2i}}{\left(m_1 c_{p1} + m_2 c_{p2} \right) T_{2i}} \right| + \\[2ex] \dfrac{m_1 c_{p1} + m_2 c_{p2}}{m_2 c_{p2} \left(T_{1i} - T_{2i} \right)} \ln \left| \dfrac{m_1 c_{p1} T_{1i} + m_2 c_{p2} T_{2i}}{m_1 c_{p1} T_{1i} + m_2 c_{p2} T_{1i}} \right| \end{array} \right]$$

(18)

Question 7

❖❖❖

The 3 plates shown are initially at a temperature T_i each. At t=0, the uppermost plate temperature is increased to T_o. Subsequently, heat transfer among the plates begins. Assume only convection mode of heat transfer to be active. The heat capacities of each of the plates is C. The convection heat transfer coefficient between plates 1 and 2 is h, and between plates 2 and 3 is also h. Surface area of each plate (engaged in heat transfer) is A. Find the uppermost plate's temperature as a function of time. Neglect temperature gradients inside each plate.

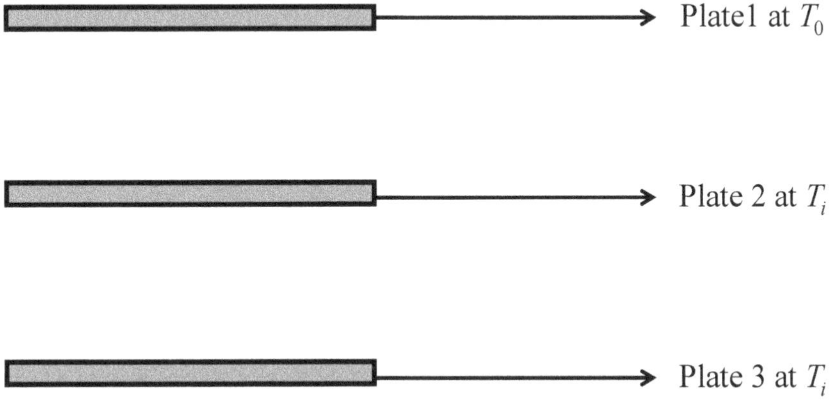

Solution

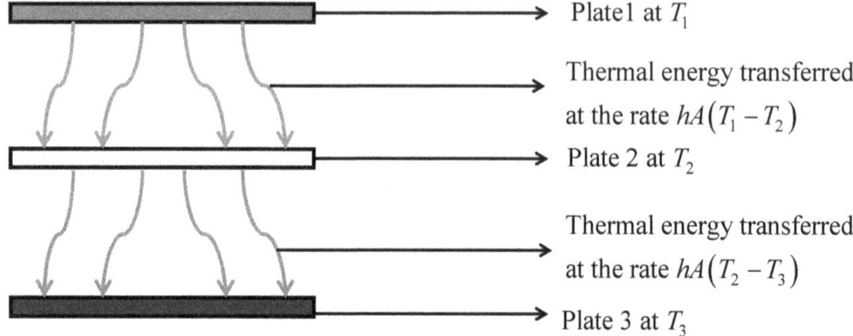

Applying law of energy conservation for the three plates at a general time t,

$$\frac{d}{dt}(CT_1) = -hA(T_1 - T_2)$$

$$\Rightarrow T_1' = -\frac{hA}{C}(T_1 - T_2) \tag{1}$$

$$\frac{d}{dt}(CT_2) = hA(T_1 - T_2) - hA(T_2 - T_3)$$

$$\Rightarrow T_2' = \frac{hA}{C}(T_1 + T_3 - 2T_2) \tag{2}$$

$$\frac{d}{dt}(CT_3) = hA(T_2 - T_3)$$

$$\Rightarrow T_3' = \frac{hA}{C}(T_2 - T_3) \tag{3}$$

From equation 1, we have

$$T_2 = T_1 + \frac{C}{hA}T_1',$$

$$\Rightarrow T_2' = T_1' + \frac{C}{hA}T_1'' \tag{4}$$

Substituting Equation 4 in 2 and 3

$$T_1' + \frac{C}{hA}T_1'' = \frac{hA}{C}\left(T_3 - T_1 - \frac{2C}{hA}T_1'\right)$$

$$\Rightarrow 3T_1' + \frac{C}{hA}T_1'' = \frac{hA}{C}(T_3 - T_1) \tag{5}$$

$$T_3' = \frac{hA}{C}\left(T_1 + \frac{C}{hA}T_1' - T_3\right) \tag{6}$$

From Equation 5, we have

$$T_3 = T_1 + \frac{3C}{hA}T_1' + \frac{C^2}{h^2 A^2}T_1''$$

$$\Rightarrow T_3' = T_1' + \frac{3C}{hA}T_1'' + \frac{C^2}{h^2 A^2}T_1''' \tag{7}$$

Substituting Equation 7 in 6

$$T_1' + \frac{3C}{hA}T_1'' + \frac{C^2}{h^2 A^2}T_1''' = -3T_1' - \frac{C}{hA}T_1'' + T_1'$$

$$\Rightarrow T_1''' + \frac{4hA}{C}T_1'' + \frac{3h^2 A^2}{C^2}T_1' = 0 \tag{8}$$

Separating the variables and integrating both sides

$$\frac{d^2 T_1}{dt^2} + \frac{4hA}{C}\frac{dT_1}{dt} + \frac{3h^2 A^2}{C^2}T_1 = C_0 \tag{9}$$

Where C_0 is an arbitrary integration constant. Equation 9 is a second order, linear, inhomogenous ordinary differential equation with constant coefficients. Its general solution can be given by method of variation of parameters as

$$T_1 = C_1 e^{s_1 t} + C_2 e^{s_2 t} + e^{s_2 t} \int \frac{e^{s_1 t} C_0 \cdot dt}{(s_2 - s_1) e^{(s_1 + s_2)t}} - e^{s_1 t} \int \frac{e^{s_2 t} C_0 \cdot dt}{(s_2 - s_1) e^{(s_1 + s_2)t}}$$

$$\Rightarrow T_1 = C_1 e^{s_1 t} + C_2 e^{s_2 t} + \frac{C_0}{(s_2 - s_1)} e^{s_2 t} \int e^{-s_2 t} \cdot dt - \frac{C_0}{(s_2 - s_1)} e^{s_1 t} \int e^{-s_1 t} \cdot dt$$

$$\Rightarrow T_1 = C_1 e^{s_1 t} + C_2 e^{s_2 t} + \frac{C_0}{(s_2 - s_1)} \left(\frac{e^{s_2 t} e^{-s_2 t}}{-s_2} \right) - \frac{C_0}{(s_2 - s_1)} \left(\frac{e^{s_1 t} e^{-s_1 t}}{-s_1} \right)$$

$$\Rightarrow T_1 = C_1 e^{s_1 t} + C_2 e^{s_2 t} + \frac{C_0}{(s_2 - s_1)} \left(\frac{1}{s_1} - \frac{1}{s_2} \right)$$

$$\Rightarrow T_1 = C_1 e^{s_1 t} + C_2 e^{s_2 t} + \frac{C_0}{s_1 s_2} \tag{10}$$

Where C_1, C_2 are integration constants and

$$s_1 = \frac{-\dfrac{4hA}{C} + \sqrt{\dfrac{16h^2 A^2}{C^2} - \dfrac{12h^2 A^2}{C^2}}}{2}, \; s_2 = \frac{-\dfrac{4hA}{C} - \sqrt{\dfrac{16h^2 A^2}{C^2} - \dfrac{12h^2 A^2}{C^2}}}{2}$$

$$\Rightarrow s_1 = -\frac{hA}{C}, \; s_2 = -\frac{3hA}{C} \tag{11}$$

Feeding equation 10 in 4 yields an expression for T_2

$$T_2 = C_1 e^{s_1 t} + C_2 e^{s_2 t} + \frac{C_0}{s_1 s_2} + \frac{C}{hA} \left(C_1 s_1 e^{s_1 t} + C_2 s_2 e^{s_2 t} \right)$$

$$\Rightarrow T_2 = C_1 e^{s_1 t} \left(1 + \frac{C s_1}{hA} \right) + C_2 e^{s_2 t} \left(1 + \frac{C s_2}{hA} \right) + \frac{C_0}{s_1 s_2} \tag{12}$$

Feeding equation 10 in 7 yields an expression for T_3

$$T_3 = C_1 e^{s_1 t} + C_2 e^{s_2 t} + \frac{C_0}{s_1 s_2} + \frac{3C}{hA} \left(C_1 s_1 e^{s_1 t} + C_2 s_2 e^{s_2 t} \right) + \frac{C^2}{h^2 A^2} \left(C_1 s_1^2 e^{s_1 t} + C_2 s_2^2 e^{s_2 t} \right)$$

$$\Rightarrow T_3 = C_1 e^{s_1 t} \left(1 + \frac{3C s_1}{hA} + \frac{C^2 s_1^2}{h^2 A^2} \right) + C_2 e^{s_2 t} \left(1 + \frac{3C s_2}{hA} + \frac{C^2 s_2^2}{h^2 A^2} \right) + \frac{C_0}{s_1 s_2} \tag{13}$$

The constants C_0, C_1, C_2 need to be evaluated using the initial conditions: $T_1(0) = T_0$, $T_2(0) = T_i$, $T_3(0) = T_i$. They yield

$$C_1 + C_2 + \frac{C_0}{s_1 s_2} = T_0 \tag{14}$$

$$C_1\left(1 + \frac{Cs_1}{hA}\right) + C_2\left(1 + \frac{Cs_2}{hA}\right) + \frac{}{s_1 s_2} = T \tag{15}$$

$$C_1\left(1 + \frac{3Cs_1}{hA} + \frac{C^2 s_1^2}{h^2 A^2}\right) + C_2\left(1 + \frac{3Cs_2}{hA} + \frac{C^2 s_2^2}{h^2 A^2}\right) + \frac{C_0}{s_1 s_2} = T_i \tag{16}$$

Solving, one gets

$$C_0 = \frac{\begin{aligned}&\left[s_1 s_2 T_0 (s_2 - s_1) C^2\right] - \left[s_2 (T_i - T_0)(hA)(2hA + Cs_2)\right] + \\ &\left[s_1 (T_i - T_0)(hA)(2hA + Cs_1)\right]\end{aligned}}{(s_2 - s_1) C^2} \tag{17}$$

$$C_1 = \frac{(T_i - T_0)(hA)(2hA + Cs_2)}{s_1 (s_2 - s_1) C^2} \tag{18}$$

$$C_2 = -\frac{(T_i - T_0)(hA)(2hA + Cs_1)}{s_2 (s_2 - s_1) C^2} \tag{19}$$

Feeding equations 17, 18, 19 in 10

$$T_1 = \frac{\left(T_i - T_0\right)\left(hA\right)\left(2hA + Cs_2\right)e^{s_1 t}}{s_1\left(s_2 - s_1\right)C^2} -$$

$$\frac{\left(T_i - T_0\right)\left(hA\right)\left(2hA + Cs_1\right)e^{s_2 t}}{s_2\left(s_2 - s_1\right)C^2} +$$

$$\frac{\left[s_1 s_2 T_0\left(s_2 - s_1\right)C^2\right] - \left[s_2\left(T_i - T_0\right)\left(hA\right)\left(2hA + Cs_2\right)\right] + \left[s_1\left(T_i - T_0\right)\left(hA\right)\left(2hA + Cs_1\right)\right]}{s_1 s_2\left(s_2 - s_1\right)C^2}$$

$$(20)$$

Expanding back s_1 and s_2

$$T_1 = \left[\frac{2T_i + T_0}{3}\right] - \left[\frac{\left(T_i - T_0\right)e^{-\frac{hAt}{C}}}{2}\right] - \left[\frac{\left(T_i - T_0\right)e^{-\frac{3hAt}{C}}}{6}\right]$$

$$(21)$$

QUESTION 8

Suppose in some parallel universe, there exists a mode of heat transfer whose controlling law is given as:

Instantaneous rate of heat transfer between 2 solid bodies $= a + b\,(\Delta T)^2$ where a,b are constants depending on the materials of the bodies and ΔT is the instantaneous temperature difference between them. If 2 bodies of masses m_1, m_2 and specific heat capacities c_{p1}, c_{p2} are allowed to exchange heat as per this law, find their temperatures as functions of time. Their initial temperatures are T_{1i}, T_{2i}. Take the former to be greater. Neglect temperature gradients inside each body.

Solution

Applying law of energy conservation for the two bodies at a general time t

$$\frac{d\left(m_1 c_{p1} T_1\right)}{dt} = -\left\{a + b\left(T_1 - T_2\right)^2\right\}$$

$$\Rightarrow m_1 c_{p1} \frac{dT_1}{dt} = -\left\{a + b\left(T_1 - T_2\right)^2\right\} \tag{1}$$

$$\frac{d\left(m_2 c_{p2} T_2\right)}{dt} = a + b\left(T_1 - T_2\right)^2$$

$$\Rightarrow m_2 c_{p2} \frac{dT_2}{dt} = a + b\left(T_1 - T_2\right)^2 \tag{2}$$

Equations 1 and 2 can be condensed as

$$-\frac{dT_1}{dt} = \alpha_1 + \beta_1\left(T_1 - T_2\right)^2 \tag{3}$$

$$\frac{dT_2}{dt} = \alpha_2 + \beta_2\left(T_1 - T_2\right)^2 \tag{4}$$

Where $\alpha_1 = \dfrac{a}{m_1 c_{p1}}, \beta_1 = \dfrac{b}{m_1 c_{p1}}, \alpha_2 = \dfrac{a}{m_2 c_{p2}}, \beta_2 = \dfrac{b}{m_2 c_{p2}}$

Equations 3 and 4 are coupled ordinary differential equations. One of the two dependent variable (T_1 and T_2) needs to be eliminated to proceed further. Using equation 4, T_1 can be expressed in terms of T_2 as

$$T_1 = T_2 \pm \sqrt{\frac{\dfrac{dT_2}{dt} - \alpha_2}{\beta_2}} \tag{5}$$

Feeding equation 5 in 3

$$-\frac{d}{dt}\left(T_2 \pm \sqrt{\frac{\frac{dT_2}{dt} - \alpha_2}{\beta_2}}\right) = \alpha_1 + \frac{\beta_1}{\beta_2}\left(\frac{dT_2}{dt} - \alpha_2\right)$$

$$\Rightarrow -\frac{dT_2}{dt} \mp \frac{d}{dt}\left(\sqrt{\frac{\frac{dT_2}{dt} - \alpha_2}{\beta_2}}\right) = \alpha_1 + \frac{\beta_1}{\beta_2}\left(\frac{dT_2}{dt} - \alpha_2\right)$$

$$\Rightarrow -\frac{dT_2}{dt} \mp \frac{1}{2\sqrt{\beta_2}\sqrt{\frac{dT_2}{dt} - \alpha_2}}\frac{d^2T_2}{dt^2} = \alpha_1 + \frac{\beta_1}{\beta_2}\left(\frac{dT_2}{dt} - \alpha_2\right)$$

$$(6)$$

Simplifying

$$\pm\frac{1}{2\sqrt{\beta_2}\sqrt{\frac{dT_2}{dt} - \alpha_2}}\frac{d^2T_2}{dt^2} + \left(1 + \frac{\beta_1}{\beta_2}\right)\frac{dT_2}{dt} = \frac{\beta_1}{\beta_2}\alpha_2 - \alpha_1$$

$$\Rightarrow \pm\frac{\frac{d^2T_2}{dt^2}}{\sqrt{\frac{dT_2}{dt} - \alpha_2}} + 2\left(\frac{\beta_1 + \beta_2}{\sqrt{\beta_2}}\right)\frac{dT_2}{dt} = 2\frac{(\beta_1\alpha_2 - \beta_2\alpha_1)}{\sqrt{\beta_2}}$$

$$(7)$$

Let $\frac{dT_2}{dt} = V$. Under this change of variable, equation 7 can then be expressed as

$$\pm\frac{\frac{dV}{dt}}{\sqrt{V - \alpha_2}} + 2\left(\frac{\beta_1 + \beta_2}{\sqrt{\beta_2}}\right)V = 2\frac{(\beta_1\alpha_2 - \beta_2\alpha_1)}{\sqrt{\beta_2}}$$

$$(8)$$

$$\pm\frac{dV}{dt\sqrt{V - \alpha_2}} = \frac{2}{\sqrt{\beta_2}}\left[\beta_1\alpha_2 - \beta_2\alpha_1 - (\beta_1 + \beta_2)V\right]$$

$$(9)$$

Separating the variables and integrating both sides

$$\pm \int \frac{dV}{\left[\beta_1\alpha_2 - \beta_2\alpha_1 - (\beta_1 + \beta_2)V\right]\sqrt{V - \alpha_2}} = \int \frac{2}{\sqrt{\beta_2}} dt + C_0 \tag{10}$$

Where C_0 is an arbitrary integration constant. Let $V - \alpha_2 = P^2 \Rightarrow dV = 2P \cdot dP$

Under this change of variable, equation 10 can be expressed as

$$\pm \int \frac{2P \cdot dP}{\left[\beta_1\alpha_2 - \beta_2\alpha_1 - \left\{(\beta_1 + \beta_2)(\alpha_2 + P^2)\right\}\right](\pm P)} = \int \frac{2}{\sqrt{\beta_2}} dt + C_0 \tag{11}$$

We can have either + or − sign outside the left hand side integral and similarly, with the term P in the denominator of this integrand. Collectively, these two sets of possibilities give rise to 2 cases , either a + or a − sign outside the integral. Simplifying further

$$\pm \int \frac{dP}{\beta_1\alpha_2 - \beta_2\alpha_1 - \left\{(\beta_1 + \beta_2)(\alpha_2 + P^2)\right\}} = \frac{t}{\sqrt{\beta_2}} + \frac{C_0}{2} \tag{12}$$

$$\mp \int \frac{dP}{(\beta_1 + \beta_2)P^2 + \beta_2\alpha_1 + \alpha_2\beta_2} = -\frac{t}{\sqrt{\beta_2}} - \frac{C_0}{2}$$

$$\Rightarrow \mp \int \frac{dP}{(\beta_1 + \beta_2)\left[P^2 + \beta_2\left(\dfrac{\alpha_1 + \alpha_2}{\beta_1 + \beta_2}\right)\right]} = -\frac{t}{\sqrt{\beta_2}} - \frac{C_0}{2} \tag{13}$$

$$\mp \int \frac{dP}{P^2 + \left(\sqrt{\dfrac{\beta_2(\alpha_1 + \alpha_2)}{\beta_1 + \beta_2}}\right)^2} = -\frac{t(\beta_1 + \beta_2)}{\sqrt{\beta_2}} - \frac{C_0(\beta_1 + \beta_2)}{2} \tag{14}$$

Using the standard result $\int \dfrac{dZ}{Z^2 + S^2} = \dfrac{1}{S} \tan^{-1}\left(\dfrac{Z}{S}\right)$

$$\mp \sqrt{\frac{\beta_1 + \beta_2}{\beta_2(\alpha_1 + \alpha_2)}}\, \tan^{-1}\left(P\sqrt{\frac{\beta_1 + \beta_2}{\beta_2(\alpha_1 + \alpha_2)}}\right) = -\frac{t(\beta_1 + \beta_2)}{\sqrt{\beta_2}} - \frac{C_0(\beta_1 + \beta_2)}{2}$$

$$\Rightarrow P = \pm \sqrt{\frac{\beta_2(\alpha_1 + \alpha_2)}{\beta_1 + \beta_2}}\, \tan\left[\left\{\frac{t(\beta_1 + \beta_2)}{\sqrt{\beta_2}} + \frac{C_0(\beta_1 + \beta_2)}{2}\right\}\sqrt{\frac{\beta_2(\alpha_1 + \alpha_2)}{\beta_1 + \beta_2}}\right] \quad (15)$$

Using the definition of P

$$V - \alpha_2 = \frac{\beta_2(\alpha_1 + \alpha_2)}{\beta_1 + \beta_2}\, \tan^2\left\{\begin{array}{l} t\sqrt{(\beta_1 + \beta_2)(\alpha_1 + \alpha_2)} + \\[2mm] \dfrac{C_0}{2}\sqrt{\beta_2(\beta_1 + \beta_2)(\alpha_1 + \alpha_2)} \end{array}\right\} \quad (16)$$

Using the definition of V

$$\frac{dT_2}{dt} = \alpha_2 +$$

$$\frac{\beta_2(\alpha_1 + \alpha_2)}{\beta_1 + \beta_2}\, \tan^2\left\{\begin{array}{l} t\sqrt{(\beta_1 + \beta_2)(\alpha_1 + \alpha_2)} + \\[2mm] \dfrac{C_0}{2}\sqrt{\beta_2(\beta_1 + \beta_2)(\alpha_1 + \alpha_2)} \end{array}\right\} \quad (17)$$

Separating the variables and integrating both sides

$$\int_{T_{2i}}^{T_2} dT_2 = \alpha_2 \int_0^t dt + \frac{\beta_2(\alpha_1 + \alpha_2)}{\beta_1 + \beta_2} \int_0^t \tan^2\left\{\begin{array}{l} t\sqrt{(\beta_1 + \beta_2)(\alpha_1 + \alpha_2)} + \\[2mm] \dfrac{C_0}{2}\sqrt{\beta_2(\beta_1 + \beta_2)(\alpha_1 + \alpha_2)} \end{array}\right\} \cdot dt \quad (18)$$

Simplifying

$$T_2 = T_{2i} + \alpha_2 t + \frac{\beta_2(\alpha_1 + \alpha_2)}{\beta_1 + \beta_2}\left[\int_0^t \sec^2\left\{\frac{t\sqrt{(\beta_1 + \beta_2)(\alpha_1 + \alpha_2)} + }{\frac{C_0}{2}\sqrt{\beta_2(\beta_1 + \beta_2)(\alpha_1 + \alpha_2)}}\right\} \cdot dt - \int_0^t 1 \cdot dt\right]$$

$$\Rightarrow T_2 = T_{2i} + \alpha_2 t + \frac{\beta_2(\alpha_1 + \alpha_2)}{\beta_1 + \beta_2}\left[\frac{\tan\left\{\frac{t\sqrt{(\beta_1 + \beta_2)(\alpha_1 + \alpha_2)} + }{\frac{C_0}{2}\sqrt{\beta_2(\beta_1 + \beta_2)(\alpha_1 + \alpha_2)}}\right\}}{\sqrt{(\beta_1 + \beta_2)(\alpha_1 + \alpha_2)}} - \frac{\tan\left\{\frac{C_0}{2}\sqrt{\beta_2(\beta_1 + \beta_2)(\alpha_1 + \alpha_2)}\right\}}{\sqrt{(\beta_1 + \beta_2)(\alpha_1 + \alpha_2)}} - t\right]$$

(19)

With the expression of T_2 available, that of T_1 can be obtained using equation 5

$$T_1 = T_{2i} + \alpha_2 t + \frac{\beta_2(\alpha_1 + \alpha_2)}{\beta_1 + \beta_2}\left[\frac{\tan\left\{\frac{t\sqrt{(\beta_1 + \beta_2)(\alpha_1 + \alpha_2)} + }{\frac{C_0}{2}\sqrt{\beta_2(\beta_1 + \beta_2)(\alpha_1 + \alpha_2)}}\right\}}{\sqrt{(\beta_1 + \beta_2)(\alpha_1 + \alpha_2)}} - \frac{\tan\left\{\frac{C_0}{2}\sqrt{\beta_2(\beta_1 + \beta_2)(\alpha_1 + \alpha_2)}\right\}}{\sqrt{(\beta_1 + \beta_2)(\alpha_1 + \alpha_2)}} - t\right] \pm$$

$$\sqrt{\frac{\alpha_1 + \alpha_2}{\beta_1 + \beta_2}}\tan\left\{t\sqrt{(\beta_1 + \beta_2)(\alpha_1 + \alpha_2)} + \frac{C_0}{2}\sqrt{\beta_2(\beta_1 + \beta_2)(\alpha_1 + \alpha_2)}\right\}$$

(20)

The unknown constant C_0 needs to be evaluated. For that, the initial condition $T_1(t=0) = T_{1i}$ needs to be used

$$T_{1i} = T_{2i} \mp \sqrt{\frac{\alpha_1 + \alpha_2}{\beta_1 + \beta_2}} \tan\left(\frac{C_0}{2} \sqrt{\beta_2 (\beta_1 + \beta_2)(\alpha_1 + \alpha_2)} \right)$$

$$\Rightarrow \tan\left(\frac{C_0}{2} \sqrt{\beta_2 (\alpha_1 + \alpha_2)(\beta_1 + \beta_2)} \right) = \mp \sqrt{\frac{\beta_1 + \beta_2}{\alpha_1 + \alpha_2}} (T_{1i} - T_{2i})$$

$$\Rightarrow \frac{C_0}{2} \sqrt{\beta_2 (\alpha_1 + \alpha_2)(\beta_1 + \beta_2)} = \tan^{-1}\left[\mp \sqrt{\frac{\beta_1 + \beta_2}{\alpha_1 + \alpha_2}} (T_{1i} - T_{2i}) \right]$$

$$\Rightarrow C_0 = \frac{\mp 2 \tan^{-1}\left\{ \sqrt{\dfrac{\beta_1 + \beta_2}{\alpha_1 + \alpha_2}} (T_{1i} - T_{2i}) \right\}}{\sqrt{\beta_2 (\alpha_1 + \alpha_2)(\beta_1 + \beta_2)}} \tag{21}$$

Feeding equation 21 in 20 and 19 and also using the definitions of $\alpha_1, \beta_1, \alpha_2, \beta_2$

$$T_1 = T_{2i} + \frac{at}{m_2 c_{p2}} + \frac{a}{m_2 c_{p2}} \left[\frac{\tan\left\{ \dfrac{t\sqrt{ab}\left(\dfrac{1}{m_1 c_{p1}} + \dfrac{1}{m_2 c_{p2}} \right) \mp}{\tan^{-1}\left\{ \sqrt{\dfrac{b}{a}}(T_{1i} - T_{2i}) \right\}} \right\}}{\sqrt{ab}\left(\dfrac{1}{m_1 c_{p1}} + \dfrac{1}{m_2 c_{p2}} \right)} \pm \dfrac{\dfrac{T_{1i} - T_{2i}}{a\left(\dfrac{1}{m_1 c_{p1}} + \dfrac{1}{m_2 c_{p2}} \right)} - t}{} \right] \pm$$

$$\sqrt{\frac{a}{b}} \tan\left\{ t\sqrt{ab}\left(\frac{1}{m_1 c_{p1}} + \frac{1}{m_2 c_{p2}} \right) \mp \tan^{-1}\left\{ \sqrt{\frac{b}{a}}(T_{1i} - T_{2i}) \right\} \right\} \tag{22}$$

$$T_2 = T_{2i} + \frac{at}{m_2 c_{p2}} + \frac{a}{m_2 c_{p2}} \left[\frac{\tan\left\{ t\sqrt{ab}\left(\dfrac{1}{m_1 c_{p1}} + \dfrac{1}{m_2 c_{p2}} \right) \mp \tan^{-1}\left\{ \sqrt{\dfrac{b}{a}}\left(T_{1i} - T_{2i} \right) \right\} \right\}}{\sqrt{ab}\left(\dfrac{1}{m_1 c_{p1}} + \dfrac{1}{m_2 c_{p2}} \right)} \pm \frac{T_{1i} - T_{2i}}{a\left(\dfrac{1}{m_1 c_{p1}} + \dfrac{1}{m_2 c_{p2}} \right)} - t \right]$$

$$\tag{23}$$

Question 9

Repeat the previous question for the heat transfer law: Instantaneous rate of heat transfer between 2 solid bodies $= b\left(T_1^2 - T_2^2\right)$ where T_1 and T_2 are their instantaneous temperatures $(T_1 > T_2)$ and b is a constant depending on the materials of the two bodies. Further, proceed to calculate the entropy generated due to the heat transfer. Consider the special case: $m_1 = m_2 = m, c_{p1} = c_{p2} = c_p$

Solution

Applying law of energy conservation for the two bodies at a general time t

$$\frac{d\left(mc_p T_1\right)}{dt} = -b\left(T_1^2 - T_2^2\right)$$

$$\Rightarrow \frac{dT_1}{dt} = -\frac{b}{mc_p}\left(T_1^2 - T_2^2\right)$$

$$\tag{1}$$

$$\frac{d\left(mc_p T_2\right)}{dt} = b\left(T_1^2 - T_2^2\right)$$

$$\Rightarrow \frac{dT_2}{dt} = \frac{b}{mc_p}\left(T_1^2 - T_2^2\right) \tag{2}$$

Equations 1 and 2 are coupled ordinary differential equations. One of the two dependent variable (T_1 and T_2) needs to be eliminated to proceed further. Dividing these 2 equations

$$\frac{dT_1}{dT_2} = -1$$

$$\Rightarrow dT_1 = -dT_2 \tag{3}$$

Integrating both sides

$$\int dT_1 = \int -dT_2$$

$$\Rightarrow T_1 = -T_2 + C_0 \quad \text{or } T_1 + T_2 = C_0 \tag{4}$$

Where C_0 is an arbitrary integration constant. Feeding T_1 from equation 4 in 2

$$\frac{dT_2}{dt} = \frac{b}{mc_p}\left(T_1 - T_2\right)\left(T_1 + T_2\right)$$

$$\Rightarrow \frac{dT_2}{dt} = \frac{bC_0}{mc_p}\left(C_0 - 2T_2\right) \tag{5}$$

Separating the variables and integrating both sides

$$\int_{T_{2i}}^{T_2} \frac{dT_2}{C_0 - 2T_2} = \int_0^t \frac{bC_0}{mc_p} dt \tag{6}$$

$$\frac{\ln\left|\dfrac{C_0 - 2T_2}{C_0 - 2T_{2i}}\right|}{-2} = \frac{bC_0 t}{mc_p} \tag{7}$$

$$\frac{C_0 - 2T_2}{C_0 - 2T_{2i}} = e^{-\frac{2bC_0 t}{mc_p}}$$

$$\Rightarrow C_0 - 2T_2 = \left(C_0 - 2T_{2i}\right)e^{-\frac{2bC_0 t}{mc_p}}$$

$$\Rightarrow T_2 = \frac{C_0}{2} - \left(\frac{C_0}{2} - T_{2i}\right)e^{-\frac{2bC_0 t}{mc_p}} \tag{8}$$

Having obtained an expression for $T_2(t)$, the same for $T_1(t)$ can be obtained by feeding equation 8 in 4

$$T_1 = C_0 - T_2$$

$$\Rightarrow T_1 = \frac{C_0}{2} + \left(\frac{C_0}{2} - T_{2i}\right)e^{-\frac{2bC_0 t}{mc_p}} \tag{9}$$

The constant C_0 needs to be evaluated using the initial condition: $T_1(0) = T_{1i}$

$$T_{1i} = \frac{C_0}{2} + \frac{C_0}{2} - T_{2i}$$

$$\Rightarrow C_0 = T_{1i} + T_{2i} \tag{10}$$

The instantaneous rate of entropy production is given by

$$\dot{S} = -\frac{\dot{Q}}{T_1} + \frac{\dot{Q}}{T_2} = \dot{Q}\left(-\frac{1}{T_1} + \frac{1}{T_2}\right) = \dot{Q}\left(\frac{T_1 - T_2}{T_1 T_2}\right) \tag{11}$$

Using the mentioned expression of instantaneous heat transfer rate

$$\dot{S} = b\left(T_1^2 - T_2^2\right)\left(\frac{T_1 - T_2}{T_1 T_2}\right) = \frac{b\left(T_1 - T_2\right)^2 \left(T_1 + T_2\right)}{T_1 T_2} \tag{12}$$

Feeding T_1 from equation 4 in 12

$$\dot{S} = \frac{bC_0\left(C_0 - 2T_2\right)^2}{\left(C_0 - T_2\right)T_2} \tag{13}$$

Feeding T_2 from equation 8 in 13

$$\dot{S} = \frac{bC_0\left(C_0 - 2T_{2i}\right)^2 e^{-\frac{4bC_0 t}{mc_p}}}{\left\{\frac{C_0}{2} - \left(\frac{C_0}{2} - T_{2i}\right)e^{-\frac{2bC_0 t}{mc_p}}\right\} \cdot \left\{\frac{C_0}{2} + \left(\frac{C_0}{2} - T_{2i}\right)e^{-\frac{2bC_0 t}{mc_p}}\right\}} \tag{14}$$

Simplifying

$$\dot{S} = \frac{4bC_0\left(C_0 - 2T_{2i}\right)^2 e^{-\frac{4bC_0 t}{mc_p}}}{\left\{C_0^2\right\} - \left\{\left(C_0 - 2T_{2i}\right)^2 e^{-\frac{4bC_0 t}{mc_p}}\right\}} \tag{15}$$

To find the net entropy generated, the above expression needs to integrated with respect to time from t=0 upto the time when heat transfer stops, which is when T_1-T_2=0. From equations 8 and 9, an expression for T_1-T_2 can be obtained and it can be seen that this quantity attains 0 at $t \to \infty$. Thus,

$$S = \int_0^\infty \frac{dS}{dt} \cdot dt = \int_0^\infty \frac{4bC_0\left(C_0 - 2T_{2i}\right)^2 e^{-\frac{4bC_0 t}{mc_p}}}{\left\{C_0^2\right\} - \left\{\left(C_0 - 2T_{2i}\right)^2 e^{-\frac{4bC_0 t}{mc_p}}\right\}} \cdot dt \tag{16}$$

Simplifying

$$S = 4bC_0\left(C_0 - 2T_{2i}\right)^2 \int_0^{\infty} \frac{dt}{\left\{C_0^2 e^{\frac{4bC_0 t}{mc_p}}\right\} - \left\{\left(C_0 - 2T_{2i}\right)^2\right\}}$$

(17)

$$\text{Let } e^{\frac{4bC_0 t}{mc_p}} = V \Rightarrow \frac{4bC_0}{mc_p}e^{\frac{4bC_0 t}{mc_p}}\,dt = dV \text{ or } dt = \frac{mc_p}{4bC_0}\frac{dV}{V}$$

$$S = mc_p\left(C_0 - 2T_{2i}\right)^2 \int_1^{\infty} \frac{dV}{V\left\{C_0^2 V - \left(C_0 - 2T_{2i}\right)^2\right\}}$$

(18)

$$S = mc_p\left(1 - \frac{2T_{2i}}{C_0}\right)^2 \int_1^{\infty} \frac{dV}{V\left\{V - \left(1 - \frac{2T_{2i}}{C_0}\right)^2\right\}}$$

(19)

$$S = \frac{mc_p\left(1 - \frac{2T_{2i}}{C_0}\right)^2}{\left(1 - \frac{2T_{2i}}{C_0}\right)^2} \int_1^{\infty} \left\{\frac{1}{\left\{V - \left(1 - \frac{2T_{2i}}{C_0}\right)^2\right\}} - \frac{1}{V}\right\} \cdot dV$$

$$\Rightarrow S = mc_p\left[\int_1^{\infty} \frac{dV}{\left\{V - \left(1 - \frac{2T_{2i}}{C_0}\right)^2\right\}} - \int_1^{\infty} \frac{dV}{V}\right]$$

(20)

$$S = mc_p \left[\ln \left| 1 - \frac{\left(1 - \frac{2T_{2i}}{C_0}\right)^2}{V} \right| \right]_1^\infty = -mc_p \ln \left| 1 - \left(1 - \frac{2T_{2i}}{C_0}\right)^2 \right| \tag{21}$$

Feeding C_0 from equation 10 in 21

$$S = -mc_p \ln \left| 1 - \left(1 - \frac{2T_{2i}}{T_{1i} + T_{2i}}\right)^2 \right|$$

$$\Rightarrow S = 2mc_p \ln \left| \frac{T_{1i} + T_{2i}}{2\sqrt{T_{1i}T_{2i}}} \right| \tag{22}$$

The final expressions for $T_1(t)$ and $T_2(t)$ can be obtained by expanding back C_0 in the equations 8 and 9

$$T_2 = \left\{ \frac{T_{1i} + T_{2i}}{2} \right\} - \left\{ \left(\frac{T_{1i} - T_{2i}}{2} \right) e^{-\frac{2b(T_{1i} + T_{2i})t}{mc_p}} \right\} \tag{23}$$

$$T_1 = \left\{ \frac{T_{1i} + T_{2i}}{2} \right\} + \left\{ \left(\frac{T_{1i} - T_{2i}}{2} \right) e^{-\frac{2b(T_{1i} + T_{2i})t}{mc_p}} \right\} \tag{24}$$

Question 10

Asolid body of mass m and specific heat capacity c_p receives heat energy at a rate proportional to the square of its instantaneous temperature and loses heat energy at a rate proportional to the square of instantaneous time derivative of its temperature [assume the body to be in some parallel universe with different heat transfer laws]. The respective proportionality constants are a and b. The initial body temperature is T_i. If the temperature is given to increase with time, find this increasing function. Neglect temperature gradients inside the body.

Solution

Applying law of energy conservation for the body at a general time t

$$\frac{d\left(mc_p T\right)}{dt} = aT^2 - b\left(\frac{dT}{dt}\right)^2$$

$$\Rightarrow mc_p \frac{dT}{dt} = aT^2 - b\left(\frac{dT}{dt}\right)^2 \tag{1}$$

Rearranging

$$b\left(\frac{dT}{dt}\right)^2 + mc_p\frac{dT}{dt} - aT^2 = 0 \tag{2}$$

Equation 2 is a quadratic in $\dfrac{dT}{dt}$. Therefore,

$$\frac{dT}{dt} = \frac{-mc_p \pm \sqrt{m^2c_p^2 + 4abT^2}}{2b} \tag{3}$$

Since temperature is given to increase with time, therefore $\dfrac{dT}{dt}$ must be positive. So, + sign is considered on the right-hand side of equation 3. Separating the variables and integrating both sides

$$\int_{T_i}^{T}\frac{dT}{\sqrt{m^2c_p^2 + 4abT^2} - mc_p} = \int_{0}^{t}\frac{dt}{2b} \tag{4}$$

Let $m^2c_p^2 + 4abT^2 = p^2 \Rightarrow 8abT\cdot dT = 2p\cdot dp$. This transforms equation 4 to

$$\int_{\sqrt{m^2c_p^2 + 4abT_i^2}}^{\sqrt{m^2c_p^2 + 4abT^2}}\frac{p\cdot dp}{4ab\sqrt{\dfrac{p^2 - m^2c_p^2}{4ab}}\left(p - mc_p\right)} = \int_{0}^{t}\frac{dt}{2b} \tag{5}$$

Simplifying

$$\int_{\sqrt{m^2c_p^2 + 4abT_i^2}}^{\sqrt{m^2c_p^2 + 4abT^2}}\frac{p\cdot dp}{\left(p - mc_p\right)\sqrt{p^2 - m^2c_p^2}} = t\sqrt{\frac{a}{b}} \tag{6}$$

$$\int\limits_{\sqrt{m^2c_p^2+4abT_i^2}}^{\sqrt{m^2c_p^2+4abT^2}} \frac{\left(p-mc_p+mc_p\right)\cdot dp}{\left(p-mc_p\right)\sqrt{p^2-m^2c_p^2}} = t\sqrt{\frac{a}{b}} \tag{7}$$

$$\int\limits_{\sqrt{m^2c_p^2+4abT_i^2}}^{\sqrt{m^2c_p^2+4abT^2}} \frac{\left(p-mc_p\right)\cdot dp}{\left(p-mc_p\right)\sqrt{p^2-m^2c_p^2}} + \int\limits_{\sqrt{m^2c_p^2+4abT_i^2}}^{\sqrt{m^2c_p^2+4abT^2}} \frac{mc_p\cdot dp}{\left(p-mc_p\right)\sqrt{p^2-m^2c_p^2}} = t\sqrt{\frac{a}{b}}$$

$$\Rightarrow \int\limits_{\sqrt{m^2c_p^2+4abT_i^2}}^{\sqrt{m^2c_p^2+4abT^2}} \frac{dp}{\sqrt{p^2-m^2c_p^2}} + mc_p \int\limits_{\sqrt{m^2c_p^2+4abT_i^2}}^{\sqrt{m^2c_p^2+4abT^2}} \frac{dp}{\left(p-mc_p\right)\sqrt{p^2-m^2c_p^2}} = t\sqrt{\frac{a}{b}} \tag{8}$$

2 integrals appear on the left hand side of equation 8. The first one can be computed using the standard result

$$\int \frac{dZ}{\sqrt{Z^2-S^2}} = \ln\left|Z+\sqrt{Z^2-S^2}\right|$$

$$\left[\ln\left|p+\sqrt{p^2-m^2c_p^2}\right|\right]_{\sqrt{m^2c_p^2+4abT_i^2}}^{\sqrt{m^2c_p^2+4abT^2}} + mc_p \int\limits_{\sqrt{m^2c_p^2+4abT_i^2}}^{\sqrt{m^2c_p^2+4abT^2}} \frac{dp}{\left(p-mc_p\right)\sqrt{p^2-m^2c_p^2}} = t\sqrt{\frac{a}{b}}$$

$$\Rightarrow \ln\left|\frac{\sqrt{m^2c_p^2+4abT^2}+2T\sqrt{ab}}{\sqrt{m^2c_p^2+4abT_i^2}+2T_i\sqrt{ab}}\right| + mc_p \int\limits_{\sqrt{m^2c_p^2+4abT_i^2}}^{\sqrt{m^2c_p^2+4abT^2}} \frac{dp}{\left(p-mc_p\right)\sqrt{p^2-m^2c_p^2}} = t\sqrt{\frac{a}{b}} \tag{9}$$

Let the other integral be named Int. It can be evaluated as shown below. First, let $p-mc_p = \dfrac{1}{v} \Rightarrow dp = -\dfrac{dv}{v^2}$

$$\text{Int} = \int\limits_{\frac{1}{\sqrt{m^2c_p^2+4abT_i^2}-mc_p}}^{\frac{1}{\sqrt{m^2c_p^2+4abT^2}-mc_p}} \frac{-\left(\dfrac{dv}{v}\right)}{\sqrt{\dfrac{1}{v^2}+\dfrac{2mc_p}{v}}} \tag{10}$$

Simplifying

$$\text{Int} = -\int_{\frac{1}{\sqrt{m^2 c_p^2 + 4abT_i^2} - mc_p}}^{\frac{1}{\sqrt{m^2 c_p^2 + 4abT^2} - mc_p}} \frac{dv}{\sqrt{1 + 2mc_p v}} \tag{11}$$

$$\text{Int} = -\left[\frac{\sqrt{1 + 2mc_p v}}{mc_p} \right]_{\frac{1}{\sqrt{m^2 c_p^2 + 4abT_i^2} - mc_p}}^{\frac{1}{\sqrt{m^2 c_p^2 + 4abT^2} - mc_p}}$$

$$\Rightarrow \text{Int} = \frac{\sqrt{1 + \dfrac{2mc_p}{\sqrt{m^2 c_p^2 + 4abT_i^2} - mc_p}}}{mc_p} - \frac{\sqrt{1 + \dfrac{2mc_p}{\sqrt{m^2 c_p^2 + 4abT^2} - mc_p}}}{mc_p}$$

$$\Rightarrow \text{Int} = \frac{\sqrt{\dfrac{\sqrt{m^2 c_p^2 + 4abT_i^2} + mc_p}{\sqrt{m^2 c_p^2 + 4abT_i^2} - mc_p}} - \sqrt{\dfrac{\sqrt{m^2 c_p^2 + 4abT^2} + mc_p}{\sqrt{m^2 c_p^2 + 4abT^2} - mc_p}}}{mc_p} \tag{12}$$

Feeding equation 12 in 9

$$\ln\left| \frac{\sqrt{m^2 c_p^2 + 4abT^2} + 2T\sqrt{ab}}{\sqrt{m^2 c_p^2 + 4abT_i^2} + 2T_i\sqrt{ab}} \right| + \sqrt{\frac{\sqrt{m^2 c_p^2 + 4abT_i^2} + mc_p}{\sqrt{m^2 c_p^2 + 4abT_i^2} - mc_p}} -$$

$$\sqrt{\frac{\sqrt{m^2 c_p^2 + 4abT^2} + mc_p}{\sqrt{m^2 c_p^2 + 4abT^2} - mc_p}} = t\sqrt{\frac{a}{b}} \tag{13}$$

Question 11

A cylinder of height H, cross-section area A is made of a material having temperature dependent thermal conductivity $k = a + bT$, where T is in kelvins and a, b are constants. Under steady state, one end of the cylinder is maintained a temperature T_0 Kelvins. The curved surface is perfectly insulated while the other end loses heat to ambience governed by anoverall heat transfer coefficient U. Find temperature of the other end if the ambient temperature is T_a Kelvins [assumed constant because of large size of ambience]. Further, find the entropy generation rate under the special case: $a=0$, $T_a=0$

Solution

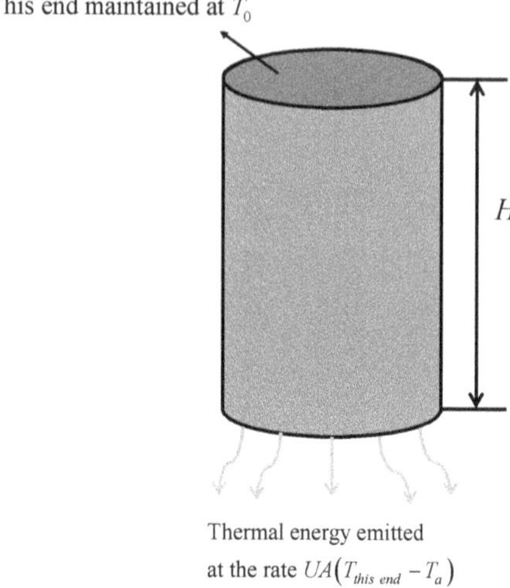

Applying law of energy conservation for the differential element shown

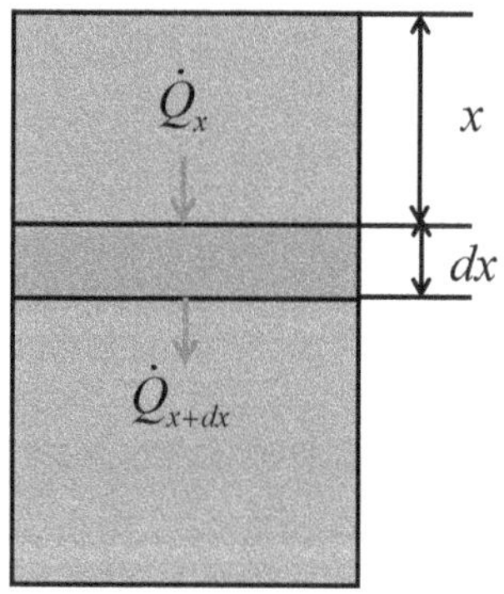

$$\dot{Q}_x = \dot{Q}_{x+dx}$$

$$\Rightarrow \dot{Q}_x = \dot{Q}_x + d\dot{Q}_x$$

$$\Rightarrow d\dot{Q}_x = 0$$

$$\Rightarrow \dot{Q}_x = \text{constant, say } C_1 \text{(a positive constant)}$$

$$\Rightarrow -kA\frac{dT}{dx} = C_1$$

(1)

Using given expression of k

$$-(a+bT)A\frac{dT}{dx} = C_1 \tag{2}$$

Separating the variables and integrating both sides

$$\int_{T_0}^{T}(a+bT)\,dT = -\int_{0}^{x}\frac{C_1}{A}\cdot dx$$

$$\Rightarrow \left[aT + b\frac{T^2}{2}\right]_{T_0}^{T} = -\frac{C_1 x}{A}$$

$$\Rightarrow a(T-T_0) + \frac{b}{2}\left(T^2 - T_0^2\right) = -\frac{C_1 x}{A}$$

$$\Rightarrow \frac{b}{2}T^2 + aT - \left(aT_0 + \frac{b}{2}T_0^2 - \frac{C_1 x}{A}\right) = 0 \tag{3}$$

Equation 3 is a quadratic in T. Therefore,

$$T = \frac{-a \pm \sqrt{a^2 + 2b\left(aT_0 + \frac{b}{2}T_0^2 - \frac{C_1 x}{A}\right)}}{b}$$

$$\Rightarrow T = \frac{-a \pm \sqrt{(a+bT_0)^2 - \frac{2bC_1 x}{A}}}{b} \tag{4}$$

Since temperature in Kelvins cannot be negative, hence + sign has to be considered in above equation. Differentiating with respect to x

$$\frac{dT}{dx} = -\frac{C_1}{A\sqrt{(a+bT_0)^2 - \frac{2bC_1 x}{A}}} \tag{5}$$

The constant C_1 has to be evaluated using the boundary condition:

$$\left(-kA\frac{dT}{dx}\right)_{x=H} = \left\{UA(T-T_a)\right\}_{x=H}$$

$$\Rightarrow -(a+bT_{x=H})\left(\frac{dT}{dx}\right)_{x=H} = U\left(T_{x=H}-T_a\right)$$

(6)

Feeding Equations 4 and 5 in 6

$$\frac{C_1}{A} = U\left(\frac{-a+\sqrt{(a+bT_0)^2 - \dfrac{2bC_1H}{A}}}{b} - T_a\right)$$

$$\Rightarrow \frac{bC_1}{UA} + bT_a + a = \sqrt{(a+bT_0)^2 - \frac{2bC_1H}{A}}$$

$$\Rightarrow \left(\frac{bC_1}{UA} + bT_a + a\right)^2 = (a+bT_0)^2 - \frac{2bC_1H}{A}$$

$$\Rightarrow \frac{b^2C_1^2}{U^2A^2} + \frac{2bC_1}{A}\left(\frac{a+bT_a+UH}{U}\right) + (a+bT_a)^2 - (a+bT_0)^2 = 0$$

(7)

Equation 7 is a quadratic in C_1. Therefore,

$$C_1 = \frac{-\dfrac{2b}{A}\left(\dfrac{a+bT_a+UH}{U}\right) \pm \sqrt{\dfrac{4b^2(a+bT_a+UH)^2}{U^2A^2} - \dfrac{4b^2\left\{(a+bT_a)^2 - (a+bT_0)^2\right\}}{U^2A^2}}}{\dfrac{2b^2}{U^2A^2}}$$

$$\Rightarrow C_1 = \frac{UA}{b}\left[-(a+bT_a+UH) \pm \sqrt{(a+bT_a+UH)^2 - \left\{(a+bT_a)^2 - (a+bT_0)^2\right\}}\right]$$

(8)

Since C_1 is local heat flow rate (equation 1), it cannot be negative. Hence, + sign needs to be considered in above equation.

The value of temperature at $x=H$ can be given by feeding equation 8 in 4 and then putting $x = H$ in the expression so obtained.

$$T_{x=H} = \frac{-a + \sqrt{\left(a+bT_0\right)^2 - \frac{2bHU}{b}\left[\begin{array}{c} -\left(a+bT_a+UH\right)+ \\ \sqrt{\left(a+bT_a+UH\right)^2 - \left\{\left(a+bT_a\right)^2 - \left(a+bT_0\right)^2\right\}} \end{array}\right]}}{b} \tag{9}$$

Entropy generation rate associated with conduction heat transfer across a differential layer is given by

$$d\dot{S} = -\frac{d\dot{Q}}{T} + \frac{d\dot{Q}}{T+dT} = d\dot{Q}\left(-\frac{1}{T} + \frac{1}{T+dT}\right) = d\dot{Q}\left(-\frac{dT}{T\left(T+dT\right)}\right) \tag{10}$$

Since dT is very small, so $T + dT \approx T$

$$d\dot{S} = -kA\frac{dT}{dx}\left(-\frac{dT}{T^2}\right) = kA\frac{dT}{dx}\left(\frac{dT}{T^2 dx}\right)dx = kA\left(\frac{\frac{dT}{dx}}{T}\right)^2 dx \tag{11}$$

The overall entropy generation rate will be integral of above

$$\dot{S} = \int d\dot{S} = A\int_0^H k\left(\frac{\frac{dT}{dx}}{T}\right)^2 dx \tag{12}$$

For the given case: $a=0$, $T_a=0$, the functions $T(x)$ and $\frac{dT}{dx}(x)$ become

$$T = \frac{\sqrt{b^2 T_0^2 - \frac{2bC_1 x}{A}}}{b} \tag{13}$$

$$\frac{dT}{dx} = -\frac{C_1}{A\sqrt{b^2 T_0^2 - \dfrac{2bC_1 x}{A}}}$$

$$(14)$$

And k becomes

$$k = bT = \sqrt{b^2 T_0^2 - \frac{2bC_1 x}{A}}$$

$$(15)$$

Feeding equations 13, 14, 15 in 12

$$\dot{S} = \frac{b^2 C_1^2}{A} \int_0^H \left(b^2 T_0^2 - \frac{2bC_1 x}{A} \right)^{-\frac{3}{2}} dx$$

$$\Rightarrow \dot{S} = -\frac{\dfrac{2b^2 C_1^2}{A}}{-\dfrac{2bC_1}{A}} \left[\left(b^2 T_0^2 - \frac{2bC_1 x}{A} \right)^{-\frac{1}{2}} \right]_0^H$$

$$\Rightarrow \dot{S} = bC_1 \left[\frac{1}{\sqrt{b^2 T_0^2 - \dfrac{2bC_1 H}{A}}} - \frac{1}{bT_0} \right]$$

$$(16)$$

The expression for C_1 for the given case can be given by equation 8

$$C_1 = \frac{UA}{b} \left[\sqrt{(UH)^2 + (bT_0)^2} - UH \right]$$

$$(17)$$

Feeding Equation 17 in 16

$$\dot{S} = UA \left[\sqrt{(UH)^2 + (bT_0)^2} - UH \right] \left[\frac{1}{\sqrt{b^2 T_0^2 - 2UH \left[\sqrt{(UH)^2 + (bT_0)^2} - UH \right]}} - \frac{1}{bT_0} \right]$$

$$(18)$$

Question 12

A frustum of end radii R_1 and R_2 $(R_2 > R_1)$ is initially at the same temperature throughout. At $t=0$, the smaller end is exposed to radiation of constant intensity I_0. The surroundings are large and their temperature is T_a. Thermal conductivity of frustum material is k. The curved surface is insulated. The overall heat transfer coefficient between each end and surroundings is U. When steady state is attained, the smaller end's temperature is T_0. Find height of frustum. Further, find entropy generation rate under the special case: $T_a = 0\ K$.

Solution

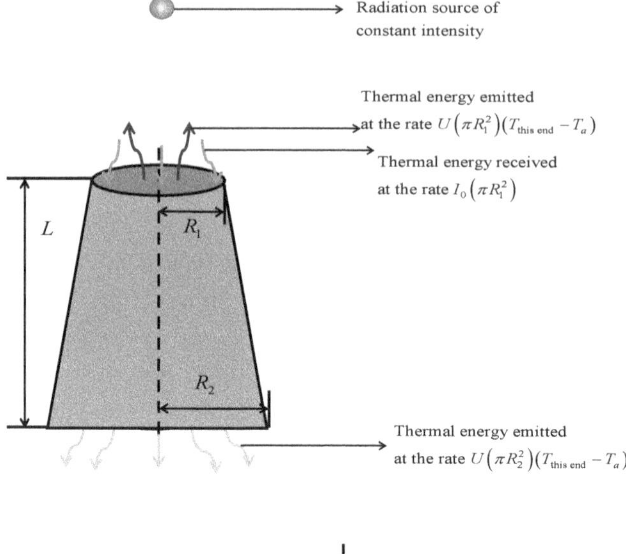

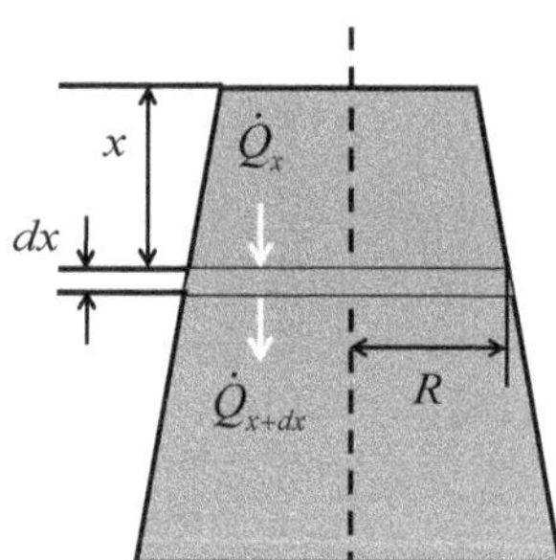

Applying law of energy conservation for the differential element shown

$$\dot{Q}_x = \dot{Q}_{x+dx}$$

$$\Rightarrow \dot{Q}_x = \dot{Q}_x + d\dot{Q}_x$$

$$\Rightarrow d\dot{Q}_x = 0$$

$$\Rightarrow \dot{Q}_x = \text{constant, say } C_1 \text{ (a positive constant)}$$

$$\Rightarrow -kA\frac{dT}{dx} = C_1 \tag{1}$$

Since radius varies linearly with x, so the local cross-sectional area can be written as

$$A = \pi R^2 = \pi \left(R_1 + mx \right)^2 \tag{2}$$

Feeding equation 2 in 1

$$-k\pi \left(R_1 + mx \right)^2 \frac{dT}{dx} = C_1 \tag{3}$$

Separating the variables and integrating both sides

$$\int_{T_0}^{T} dT = \int_{0}^{x} \frac{-C_1 dx}{k\pi \left(R_1 + mx \right)^2}$$

$$\Rightarrow T - T_0 = \frac{C_1}{k\pi m} \left[\frac{1}{R_1 + mx} - \frac{1}{R_1} \right]$$

$$\Rightarrow T = T_0 - \frac{C_1 x}{k\pi R_1 \left(R_1 + mx \right)} \tag{4}$$

The relevant boundary conditions are

$$I_0 A_{x=0} = -kA_{x=0} \left(\frac{dT}{dx} \right)_{x=0} + UA_{x=0} \left(T_{x=0} - T_a \right) \tag{5}$$

$$-kA_{x=L} \left(\frac{dT}{dx} \right)_{x=L} = UA_{x=L} \left(T_{x=L} - T_a \right) \tag{6}$$

Feeding equations 3 and 4 in 5 and 6, and also using $R_1 + mL = R_2$

$$I_0 = \frac{C_1}{\pi R_1^2} + U \left(T_0 - T_a \right) \tag{7}$$

$$\frac{C_1}{\pi R_2{}^2} = U\left(T_0 - \frac{C_1 L}{k\pi R_1 R_2} - T_a\right)$$

(8)

Two equations in two unknowns, C_1 and L. Eliminating C_1 from these two equations

$$L = \frac{k\left\{U\left(R_1{}^2 + R_2{}^2\right)\left(T_0 - T_a\right) - I_0 R_1{}^2\right\}}{UR_1 R_2\left\{I_0 - U\left(T_0 - T_a\right)\right\}}$$

(9)

Entropy generation rate associated with conduction heat transfer across a differential layer is

$$d\dot{S} = -\frac{d\dot{Q}}{T} + \frac{d\dot{Q}}{T+dT} = d\dot{Q}\left(-\frac{1}{T} + \frac{1}{T+dT}\right) = d\dot{Q}\left(-\frac{dT}{T(T+dT)}\right)$$

(10)

Since dT is very small, so $T + dT \approx T$

$$d\dot{S} = -kA\frac{dT}{dx}\left(-\frac{dT}{T^2}\right) = kA\frac{dT}{dx}\left(\frac{dT}{T^2 dx}\right)dx = kA\left(\frac{\frac{dT}{dx}}{T}\right)^2 dx$$

(11)

The overall entropy generation rate will be integral of above

$$\dot{S} = \int d\dot{S} = k\int_0^L A\left(\frac{\frac{dT}{dx}}{T}\right)^2 dx$$

(12)

Feeding equations 2, 3, 4 in 12

$$\dot{S} = k\pi C_1^2 R_1^2 \int\limits_0^L \frac{dx}{\left\{T_0 k\pi R_1^2 + \left(T_0 k\pi R_1 m - C_1\right)x\right\}^2}$$

$$\Rightarrow \dot{S} = -\frac{k\pi C_1^2 R_1^2}{\left(T_0 k\pi R_1 m - C_1\right)}\left[\frac{1}{T_0 k\pi R_1^2 + \left(T_0 k\pi R_1 m - C_1\right)x}\right]_0^L$$

$$\Rightarrow \dot{S} = \frac{C_1^2 L}{T_0\left\{T_0 k\pi R_1^2 + \left(T_0 k\pi R_1 m - C_1\right)L\right\}}$$

$$\Rightarrow \dot{S} = \frac{C_1^2 L}{T_0\left\{T_0 k\pi R_1^2 + \left\{T_0 k\pi R_1\left(\dfrac{R_2 - R_1}{L}\right) - C_1\right\}L\right\}}$$

$$\Rightarrow \dot{S} = \frac{C_1^2 L}{T_0\left[T_0 k\pi R_1^2 + \left\{T_0 k\pi R_1\left(R_2 - R_1\right)L - C_1 L\right\}\right]} \tag{13}$$

Feeding equation 9 in 13 for the given case: $T_a = 0\,K$

$$\dot{S} = \frac{C_1^2\,\dfrac{k\left\{UT_0\left(R_1^2 + R_2^2\right) - I_0 R_1^2\right\}}{T_0}}{\left[T_0 k\pi R_1^3 UR_2\left(I_0 - UT_0\right)\right] + \left[\dfrac{k^2 T_0 \pi R_1\left(R_2 - R_1 - C_1\right)}{\left\{UT_0\left(R_1^2 + R_2^2\right) - I_0 R_1^2\right\}}\right]} \tag{14}$$

Feeding C_1 from equation 7 in 14 for the given case: $T_a = 0\,K$

$$\dot{S} = \frac{\dfrac{k\left\{\pi R_1^2\left(I_0 - UT_0\right)\right\}^2\left\{UT_0\left(R_1^2 + R_2^2\right) - I_0 R_1^2\right\}}{T_0}}{\left[T_0 k\pi R_1^3 UR_2\left(I_0 - UT_0\right)\right] + \left[\dfrac{k^2 T_0 \pi R_1\left(R_2 - R_1 - \pi R_1^2 I_0 + \pi R_1^2 UT_0\right)}{\left\{UT_0\left(R_1^2 + R_2^2\right) - I_0 R_1^2\right\}}\right]} \tag{15}$$

Question 13

A cylindrical rod of cross-sectional area A, length L has one end of it immersed in a large thermal reservoir at temperature T_0. The other end and the curved surface are insulated. Heat is generated inside the rod with the volumetric heat generation rate q_g being location dependent as $q_g = x^2 e^{ax}$ where x is distance measured from the non-insulated end and a is a positive constant. Find the steady state temperature distribution in the rod. Thermal conductivity of the rod material is k

Solution

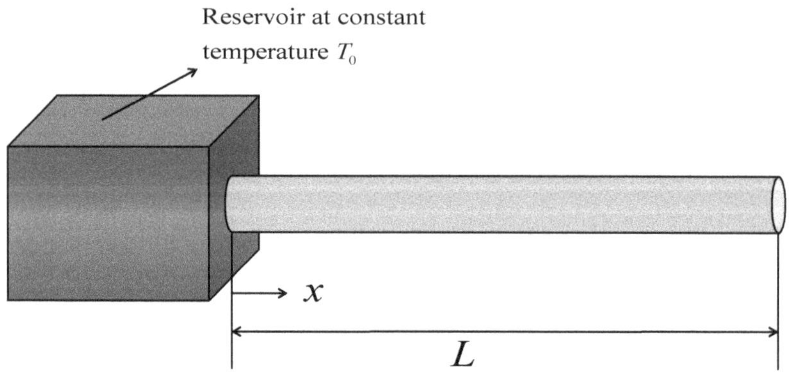

Applying law of energy conservation for the differential element shown,

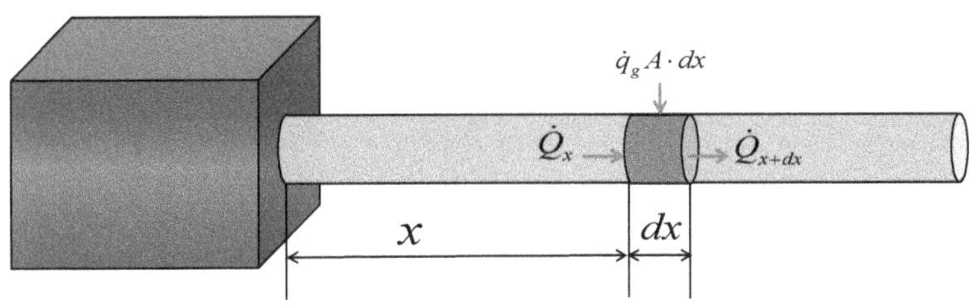

$$\dot{Q}_x - \dot{Q}_{x+dx} + \dot{q}_g \left(A \cdot dx \right) = 0$$

$$\Rightarrow \dot{Q}_x - \left(\dot{Q}_x + d\dot{Q}_x \right) + \dot{q}_g \left(A \cdot dx \right) = 0$$

$$\Rightarrow -d\dot{Q}_x + \dot{q}_g \left(A \cdot dx \right) = 0$$

$$\Rightarrow -d\left(-kA \frac{dT}{dx} \right) + \dot{q}_g \left(A \cdot dx \right) = 0$$

$$\Rightarrow kA \frac{d}{dx}\left(\frac{dT}{dx} \right) + \dot{q}_g A = 0$$

$$\Rightarrow \frac{d^2 T}{dx^2} + \frac{\dot{q}_g}{k} = 0 \qquad (1)$$

Feeding the mentioned expression of $\dot{q}_g$ in equation 1

$$\frac{d^2 T}{dx^2} + \frac{x^2 e^{ax}}{k} = 0 \qquad (2)$$

Separating the variables tand integrating both sides

$$\frac{d}{dx}\left(\frac{dT}{dx}\right)+\frac{x^2 e^{ax}}{k}=0$$

$$\Rightarrow d\left(\frac{dT}{dx}\right)+\frac{x^2 e^{ax}\cdot dx}{k}=0$$

$$\Rightarrow \int d\left(\frac{dT}{dx}\right)+\int \frac{x^2 e^{ax}\cdot dx}{k}=C_1$$

$$\Rightarrow \frac{dT}{dx}+\int \frac{x^2 e^{ax}\cdot dx}{k}=C_1 \tag{3}$$

Where C_1 is an arbitrary integration constant. The integral appearing in equation 3 can be evaluated using integration by parts as shown below.

$$\int x^2 e^{ax}\cdot dx = x^2 \int e^{ax}\cdot dx - \int 2x \int e^{ax}\cdot dx$$

$$\Rightarrow \int x^2 e^{ax}\cdot dx = x^2 \frac{e^{ax}}{a}-\frac{2}{a}\int xe^{ax}\cdot dx \tag{4}$$

Again applying integration by parts on the integral appearing on the right hand side of equation 4

$$\int x^2 e^{ax}\cdot dx = x^2 \frac{e^{ax}}{a}-\frac{2}{a}\left[x\int e^{ax}\cdot dx - \int\int e^{ax}\cdot dx\right]$$

$$\Rightarrow \int x^2 e^{ax}\cdot dx = \frac{x^2 e^{ax}}{a}-\frac{2}{a}\left[\frac{xe^{ax}}{a}-\int \frac{e^{ax}\cdot dx}{a}\right]$$

$$\Rightarrow \int x^2 e^{ax}\cdot dx = \frac{x^2 e^{ax}}{a}-\frac{2xe^{ax}}{a^2}+\frac{2e^{ax}}{a^3}$$

$$\Rightarrow \int x^2 e^{ax}\cdot dx = \frac{e^{ax}}{a}\left(x^2 -\frac{2x}{a}+\frac{2}{a^2}\right) \tag{5}$$

The required integral is now complete. Feeding it from equation 5 in 3

$$\frac{dT}{dx} + \frac{e^{ax}}{ka}\left(x^2 - \frac{2x}{a} + \frac{2}{a^2} \right) = C \tag{6}$$

Separating the variables and integrating both sides

$$dT + \frac{e^{ax}}{ka}\left(x^2 - \frac{2x}{a} + \frac{2}{a^2} \right)dx = C_1 \cdot dx$$

$$\Rightarrow \int dT + \int \frac{e^{ax}}{ka}\left(x^2 - \frac{2x}{a} + \frac{2}{a^2} \right)dx = \int C_1 \cdot dx$$

$$\Rightarrow T + \frac{\displaystyle\int e^{ax}\left(x^2 - \frac{2x}{a} + \frac{2}{a^2} \right)dx}{ka} = C_1 x + C_2 \tag{7}$$

Where C_2 is an arbitrary integration constant. The integral appearing in equation 7 can be broken into 3 integrals as shown in equation 8. The third one is a standard integral, the first one has just been evaluated and second one was evaluated as an intermediate step in the evaluation of the first one

$$T + \frac{\int x^2 e^{ax} \cdot dx}{ka} - \frac{2\int x e^{ax} \cdot dx}{ka^2} + \frac{2\int e^{ax} \cdot dx}{ka^3} = C_1 x + C_2 \tag{8}$$

$$T + \frac{e^{ax}}{ka^2}\left(x^2 - \frac{2x}{a} + \frac{2}{a^2} \right) - \frac{2e^{ax}}{ka^3}\left(x - \frac{1}{a} \right) + \frac{2e^{ax}}{ka^4} = C_1 x + C_2$$

$$\Rightarrow T + \frac{e^{ax}}{ka^2}\left(x^2 - \frac{2x}{a} + \frac{2}{a^2} - \frac{2x}{a} + \frac{2}{a^2} + \frac{2}{a^2} \right) = C_1 x + C_2$$

$$\Rightarrow T = C_1 x + C_2 - \frac{e^{ax}\left(x^2 - \frac{4x}{a} + \frac{6}{a^2} \right)}{ka^2} \tag{9}$$

The unknown constants C_1 and C_2 have to be found using the boundary conditions written ahead

$$T_{x=0} = T_0 \tag{10}$$

$$\left(\frac{dT}{dx} \right)_{x=L} = 0 \tag{11}$$

Feeding equation 9 in equations 10 and 11

$$C_2 = \frac{6}{ka^4} + T_0 \tag{12}$$

$$C_1 = \frac{e^{aL}}{ka^2} \left(aL^2 - 2L + \frac{2}{a} \right) \tag{13}$$

With the constants evaluated, the temperature distribution is now complete. Feeding equations 12 and 13 in 9

$$T = T_0 + \frac{6}{ka^4} + \frac{e^{aL} \left(aL^2 - 2L + \dfrac{2}{a} \right) x}{ka^2} - \frac{e^{ax} \left(x^2 - \dfrac{4x}{a} + \dfrac{6}{a^2} \right)}{ka^2} \tag{14}$$

Question 14

A fin is a solid extension attached to an object in order to increase its effective surface area exposed to surroundings, and therefore the heat transfer rate with them. In the conventional analysis of fins taught in standard heat transfer textbooks, the convection heat transfer coefficient is taken to be constant while radiation heat transfer is neglected. Consider a fin made of a material having thermal conductivity k, cross sectional area A and circumferential perimeter P. The convection heat transfer coefficient however, varies with distance x from the object as $h = \dfrac{h_0}{(a+bx)^2}$ where h_0, a, b are positive constants. Under steady state, if the object temperature is T_0, find the rate of heat transfer between the fin and its surroundings. The surroundings are large and at temperature T_a and fin length is L. The end of the fin exposed to surroundings is perfectly insulated. Neglect multi-dimensional heat conduction in the rod, and also radiative heat transfer between rod and surroundings.

Solution

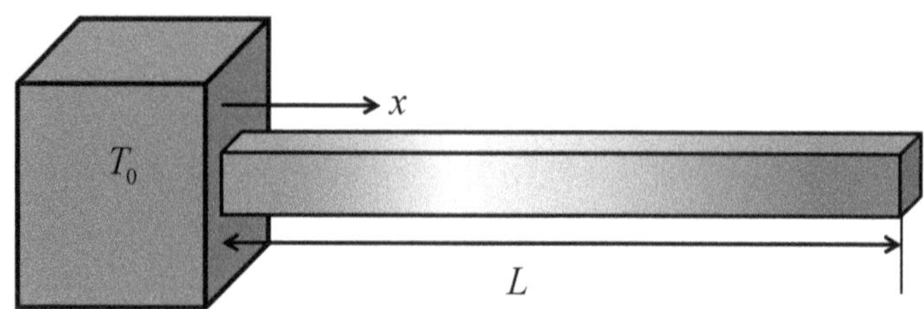

Applying law of energy conservation for the differential fin element shown

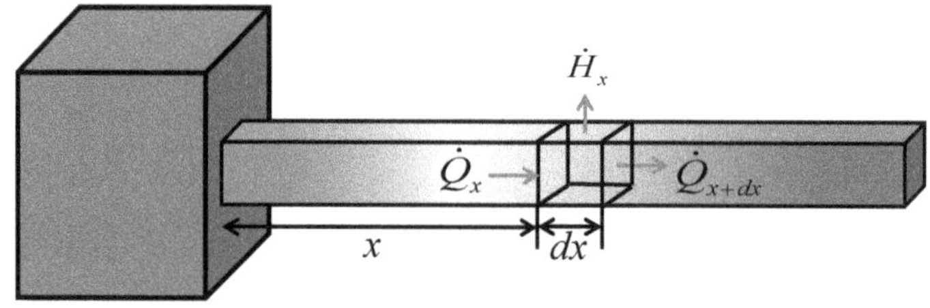

$$\dot{Q}_x - \dot{Q}_{x+dx} - \dot{H}_x = 0$$

$$\Rightarrow \dot{Q}_x - \left(\dot{Q}_x + d\dot{Q}_x\right) - \dot{H}_x = 0$$

$$\Rightarrow d\dot{Q}_x + \dot{H}_x = 0 \tag{1}$$

$$d\left(-kA\frac{dT}{dx}\right) + h(P \cdot dx)(T - T_a) = 0$$

$$\Rightarrow \frac{d}{dx}\left(-kA\frac{dT}{dx}\right) + \frac{h_0 P(T - T_a)}{(a + bx)^2} = 0$$

$$\Rightarrow \frac{d^2 T}{dx^2} = \frac{h_0 P}{kA(a + bx)^2}(T - T_a) \tag{2}$$

Let $T - T_a = \theta$. Equation 2 can then be rearranged as

$$\frac{d^2\theta}{dx^2} = \frac{h_0 P \theta}{kA(a+bx)^2}$$

(3)

Equation 3 is a 2nd order, linear, homogenous Euler Cauchy ordinary differential equation. The method to solve it is: First assume a trial solution of the form $\theta = (a+bx)^\lambda$ where $\lambda \in R$. Then Feed this in equation 3,

$$b^2 \lambda (\lambda - 1)(a+bx)^{\lambda-2}(a+bx)^2 = \frac{h_0 P (a+bx)^\lambda}{kA}$$

$$\Rightarrow \lambda^2 - \lambda - \frac{h_0 P}{kAb^2} = 0$$

(4)

Equation 4 is a quadratic in λ. Its two roots are

$$\lambda_1 = \frac{1 + \sqrt{1 + \dfrac{4h_0 P}{kAb^2}}}{2}, \lambda_2 = \frac{1 - \sqrt{1 + \dfrac{4h_0 P}{kAb^2}}}{2}$$

(5)

As per the superposition principle, the general solution $\theta(x)$ for the ODE under considerationis a linear combination of two of its particular solutions. Thus, the general solution is

$$\theta = C_1 (a+bx)^{\lambda_1} + C_2 (a+bx)^{\lambda_2}$$

(6)

Where C_1 and C_2 are arbitrary integration constants. They need to be found by the pertinent boundary conditions which are

$$T_{x=0} = T_0$$
$$\Rightarrow \theta_{x=0} = T_0 - T_a \tag{7}$$

$$-kA\left(\frac{dT}{dx}\right)_{x=L} = 0$$
$$\Rightarrow \left(\frac{dT}{dx}\right)_{x=L} = 0$$
$$\Rightarrow \left(\frac{d\theta}{dx}\right)_{x=L} = 0 \tag{8}$$

Feeding equation 6 in equations 7 and 8

$$C_1 a^{\lambda_1} + C_2 a^{\lambda_2} = T_0 - T_a \tag{9}$$

$$C_1 b\lambda_1 (a+bL)^{\lambda_1-1} + C_2 b\lambda_2 (a+bL)^{\lambda_2-1} = 0$$
$$\Rightarrow C_2 = -\frac{C_1\lambda_1 (a+bL)^{\lambda_1-\lambda_2}}{\lambda_2} \tag{10}$$

Feeding equation 10 in 9 gives an expression for C_1

$$C_1 a^{\lambda_1} - \frac{C_1\lambda_1 a^{\lambda_2} (a+bL)^{\lambda_1-\lambda_2}}{\lambda_2} = T_0 - T_a$$
$$\Rightarrow C_1 = \frac{\lambda_2 (T_0 - T_a)}{a^{\lambda_1}\lambda_2 - \lambda_1 a^{\lambda_2} (a+bL)^{\lambda_1-\lambda_2}} \tag{11}$$

Feeding equation 11 in 10 gives an expression for C_2

$$C_2 = -\frac{\lambda_2\lambda_1 (a+bL)^{\lambda_1-\lambda_2} (T_0 - T_a)}{a^{\lambda_1}\lambda_2^2 - \lambda_1\lambda_2 a^{\lambda_2} (a+bL)^{\lambda_1-\lambda_2}} \tag{12}$$

This completes the required temperature distribution. The net heat transfer rate from the fin will be the integral of the heat transfer rate from a differential fin length

$$\dot{Q}_{net} = \int_0^L h(P \cdot dx)(T - T_a) = P\int_0^L h\theta \cdot dx \tag{13}$$

Feeding equation 6 and the provided expression of h in equation 13

$$\dot{Q}_{net} = P\int_0^L \frac{h_0\left[C_1(a+bx)^{\lambda_1} + C_2(a+bx)^{\lambda_2}\right]}{(a+bx)^2} \cdot dx$$

$$\Rightarrow \dot{Q}_{net} = h_0 P\int_0^L \left[\frac{C_1(a+bx)^{\lambda_1}}{(a+bx)^2} + \frac{C_2(a+bx)^{\lambda_2}}{(a+bx)^2}\right] \cdot dx$$

$$\Rightarrow \dot{Q}_{net} = h_0 P\int_0^L \left[C_1(a+bx)^{\lambda_1-2} + C_2(a+bx)^{\lambda_2-2}\right] \cdot dx \tag{14}$$

Simplifying

$$\dot{Q}_{net} = h_0 P\left[\frac{C_1(a+bx)^{\lambda_1-1}}{b(\lambda_1-1)} + \frac{C_2(a+bx)^{\lambda_2-1}}{b(\lambda_2-1)}\right]_0^L$$

$$\Rightarrow \dot{Q}_{net} = h_0 P\left[\frac{C_1(a+bL)^{\lambda_1-1} - C_1 a^{\lambda_1-1}}{b(\lambda_1-1)} + \frac{C_2(a+bL)^{\lambda_2-1} - C_2 a^{\lambda_2-1}}{b(\lambda_2-1)}\right] \tag{15}$$

Feeding C_1 and C_2 from equations 11 and 12 in 15

$$\dot{Q}_{net} = \frac{h_0 P \lambda_2 \left(T_0 - T_a\right)}{b} \left[\frac{\left\{ \left(a+bL\right)^{\lambda_1-1} - a^{\lambda_1-1} \right\}}{\left(\lambda_1-1\right)\left[a^{\lambda_1}\lambda_2 - \lambda_1 a^{\lambda_2}\left(a+bL\right)^{\lambda_1-\lambda_2} \right]} - \frac{\lambda_1\left(a+bL\right)^{\lambda_1-\lambda_2}\left\{ \left(a+bL\right)^{\lambda_2-1} - a^{\lambda_2-1} \right\}}{\left(\lambda_2-1\right)\left[a^{\lambda_1}\lambda_2^2 - \lambda_1\lambda_2 a^{\lambda_2}\left(a+bL\right)^{\lambda_1-\lambda_2} \right]} \right] \tag{16}$$

Expanding back λ_1, λ_2

$$\dot{Q}_{net} = \frac{h_0 P}{b}\left(\frac{1-\sqrt{1+\frac{4h_0 P}{kAb^2}}}{2} \right)\left(T_0 - T_a\right) \left[\frac{\left(\frac{\sqrt{1+\frac{4h_0 P}{kAb^2}}-1}{2} \right)\left[\dfrac{\left(a+bL\right)^{\frac{\sqrt{1+\frac{4h_0 P}{kAb^2}}-1}{2}} - a^{\frac{\sqrt{1+\frac{4h_0 P}{kAb^2}}-1}{2}}}{a^{\frac{1+\sqrt{1+\frac{4h_0 P}{kAb^2}}}{2}}\left(\frac{1-\sqrt{1+\frac{4h_0 P}{kAb^2}}}{2} \right) - \left(\frac{1+\sqrt{1+\frac{4h_0 P}{kAb^2}}}{2} \right)a^{\frac{1-\sqrt{1+\frac{4h_0 P}{kAb^2}}}{2}}\left(a+bL\right)^{\sqrt{1+\frac{4h_0 P}{kAb^2}}}} \right] }{ \left(\frac{-\sqrt{1+\frac{4h_0 P}{kAb^2}}-1}{2} \right)\left[a^{\frac{1+\sqrt{1+\frac{4h_0 P}{kAb^2}}}{2}}\left(\frac{1-\sqrt{1+\frac{4h_0 P}{kAb^2}}}{2} \right)^2 - \left(\frac{1+\sqrt{1+\frac{4h_0 P}{kAb^2}}}{2} \right)\left(\frac{1-\sqrt{1+\frac{4h_0 P}{kAb^2}}}{2} \right)a^{\frac{1-\sqrt{1+\frac{4h_0 P}{kAb^2}}}{2}}\left(a+bL\right)^{\sqrt{1+\frac{4h_0 P}{kAb^2}}} \right] } \right] \tag{17}$$

Question 15

A truncated solid hemisphere is placed on the ground. Radiation from a far-off source, of intensity I_0 falls on it. The curved surface and base are insulated whereas the top surface is left bare. It receives radiative energy and also loses heat to surroundings as per the relation: $\dot{Q}_{loss} = h_{sur} A_{x=0} \left(T_{x=0} - T_a \right)$ where T_a is ambient temperature (consider it constant because of large size of surroundings.) and h_{sur} is overall heat transfer coefficient from surface to surroundings. Further, radiation decays exponentially inside the body as per the relation: $I_x = I_0 e^{-\mu x}$ where μ is a positive constant and x is distance travelled by radiation in the solid. Radius of the solid is R, its height is H and thermal conductivity of its material is k. Derive an expression for the steady state temperature distribution inside the truncated hemisphere.

Solution

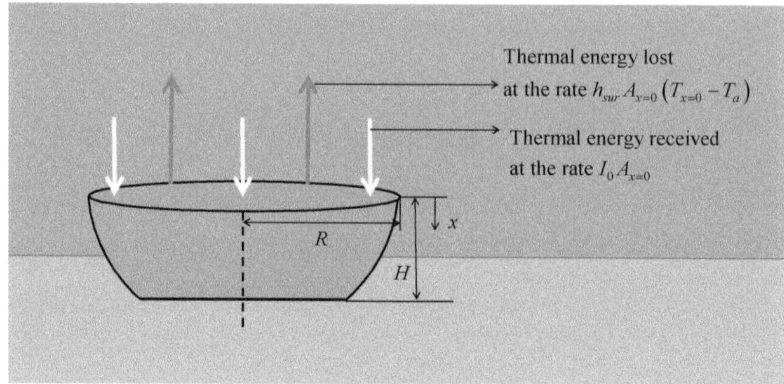

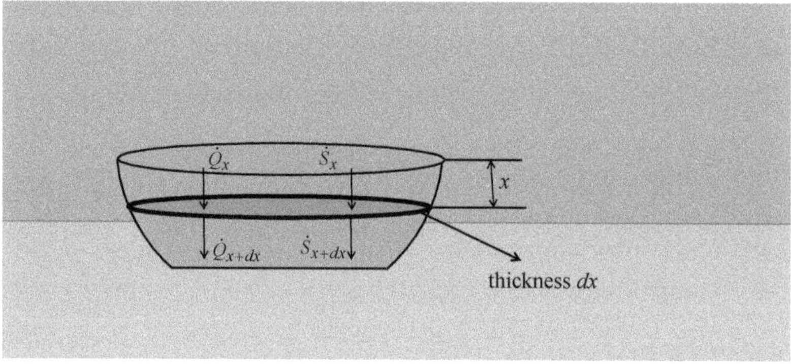

Applying law of energy conservation for the differential element shown,

$$\dot{Q}_x - \dot{Q}_{x+dx} + \dot{S}_x - \dot{S}_{x+dx} = 0$$

$$\Rightarrow \dot{Q}_x - \left(\dot{Q}_x + d\dot{Q}_x\right) + \dot{S}_x - \left(\dot{S}_x + d\dot{S}_x\right) = 0$$

$$\Rightarrow -d\dot{Q}_x - d\dot{S}_x = 0$$

$$\Rightarrow -d\left(-kA\frac{dT}{dx}\right) - d\left(I_0 Ae^{-\mu x}\right) = 0$$

$$\Rightarrow -kA\frac{dT}{dx} + I_0 Ae^{-\mu x} = C_1 \tag{1}$$

Where C_1 is an arbitrary integration constant. The cross-section area normal to heat flow is variable. Since each cross section is a circle, so local area will be πr^2 where r is the local radius. This local radius can be found using geometry, yielding an expression for local area mentioned below

$$A = \pi\left(R^2 - x^2\right)$$

(2)

Feeding equation 2 in 1

$$-k\frac{dT}{dx} + I_0 e^{-\mu x} = \frac{C_1}{\pi\left(R^2 - x^2\right)}$$

(3)

Separating the variables and integrating both sides

$$-k \cdot dT + I_0 e^{-\mu x} \cdot dx = \frac{C_1 \cdot dx}{\pi\left(R^2 - x^2\right)}$$

$$\Rightarrow \int -k \cdot dT + \int I_0 e^{-\mu x} \cdot dx = \int \frac{C_1 \cdot dx}{\pi\left(R^2 - x^2\right)}$$

$$\Rightarrow -kT - \frac{I_0 e^{-\mu x}}{\mu} = \frac{C_1}{2\pi R}\ln\left|\frac{x+R}{x-R}\right| + C_2$$

(4)

Where C_2 is an arbitrary integration constant. Simplifying

$$T = \frac{C_1}{2\pi kR}\ln\left|\frac{x-R}{x+R}\right| - \frac{C_2}{k} - \frac{I_0 e^{-\mu x}}{\mu k}$$

(5)

The constants C_1 and C_2 need to be found by the relevant boundary conditions mentioned below

$$I_0 = h_{sur}\left(T_{x=0} - T_a\right) - k\left(\frac{dT}{dx}\right)_{x=0} \tag{6}$$

$$\left(\frac{dT}{dx}\right)_{x=H} = 0 \tag{7}$$

Feeding equation 5 in equations 6 and 7

$$I_0 = h_{sur}\left(-\frac{C_2}{k} - \frac{I_0}{\mu k} - T_a\right) - k\left(\frac{I_0}{k} - \frac{C_1}{k\pi R^2}\right) \tag{8}$$

$$C_1 = I_0 \pi\left(R^2 - H^2\right)e^{-\mu H} \tag{9}$$

Feeding equation 9 in 8

$$I_0 = h_{sur}\left(-\frac{C_2}{k} - \frac{I_0}{\mu k} - T_a\right) - I_0 + \frac{I_0\left(R^2 - H^2\right)e^{-\mu H}}{R^2}$$

$$\Rightarrow C_2 = \frac{\dfrac{I_0 k}{R^2}\left(R^2 - H^2\right)e^{-\mu H} - 2I_0 k}{h_{sur}} - \frac{I_0}{\mu} - kT_a \tag{10}$$

Having found out C_1 and C_2, they are to now be fed to equation 5 to complete the required temperature distribution

$$T = \frac{I_0\left(R^2 - H^2\right)e^{-\mu H}}{2kR}\ln\left|\frac{x - R}{x + R}\right| - \frac{\dfrac{I_0}{R^2}\left(R^2 - H^2\right)e^{-\mu H} - 2I_0}{h_{sur}} + \frac{I_0}{\mu k} + T_a - \frac{I_0 e^{-\mu x}}{\mu k} \tag{11}$$

Question 16

A topic taught in heat transfer courses is one dimensional steady state conduction with uniform heat generation. 3 different geometries are generally investigated under this topic: A plane wall, a solid cylinder and a solid sphere suffering internal heat generation at a uniform (location independent) volumetric heat generation rate $\dot{q}_g$. The outer surface of the object is exposed to surroundings and loses energy to them via convection alone, wherein the convection heat transfer coefficient is U. The surroundings are large and at temperature T_a (considered constant). Thermal conductivity of the object is k. Considering one dimensional steady state heat transfer, the following analytical expressions for temperature distributions are obtained:

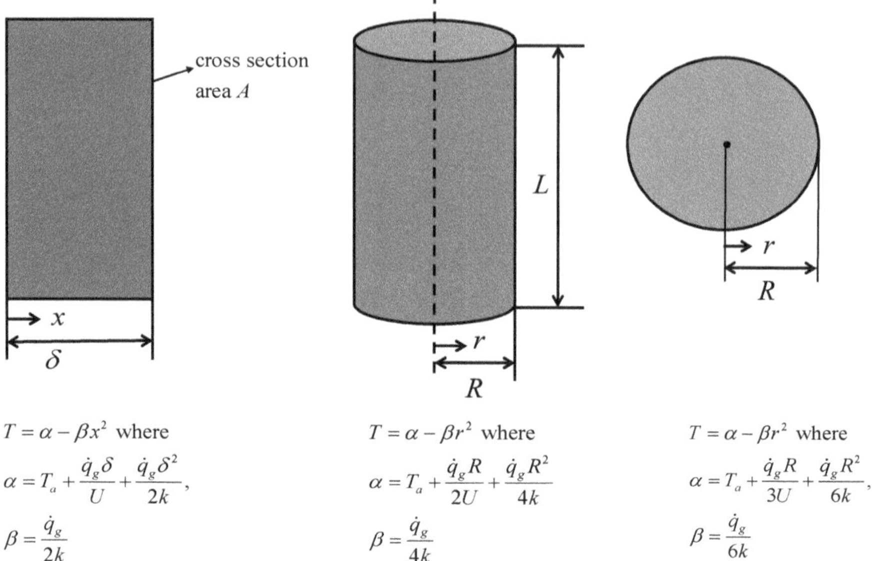

$T = \alpha - \beta x^2$ where

$$\alpha = T_a + \frac{\dot{q}_g \delta}{U} + \frac{\dot{q}_g \delta^2}{2k},$$

$$\beta = \frac{\dot{q}_g}{2k}$$

$T = \alpha - \beta r^2$ where

$$\alpha = T_a + \frac{\dot{q}_g R}{2U} + \frac{\dot{q}_g R^2}{4k}$$

$$\beta = \frac{\dot{q}_g}{4k}$$

$T = \alpha - \beta r^2$ where

$$\alpha = T_a + \frac{\dot{q}_g R}{3U} + \frac{\dot{q}_g R^2}{6k},$$

$$\beta = \frac{\dot{q}_g}{6k}$$

Find the steady state entropy generation rate in each of these cases.

Solution

CASE 1: Entropy generation rate associated with conduction heat transfer across a differential layer is

$$d\dot{S}_1 = -\frac{d\dot{Q}}{T} + \frac{d\dot{Q}}{T + dT} = d\dot{Q}\left[-\frac{1}{T} + \frac{1}{T + dT} \right] = -d\dot{Q}\left[\frac{dT}{T(T + dT)} \right] \tag{1}$$

Since dT is very small, $T + dT \approx T$

$$d\dot{S}_1 = -\left(-kA\frac{dT}{dx} \right)\left(\frac{dT}{T^2} \right) = \left(kA\frac{dT}{dx} \right)\left(\frac{dT}{T^2 dx} \right)dx = kA\left(\frac{\frac{dT}{dx}}{T} \right)^2 dx \tag{2}$$

The overall entropy generation rate due to conduction will be the integral of above

$$\dot{S}_1 = \int d\dot{S}_1 = \int_0^\delta kA \left(\frac{\frac{dT}{dx}}{T} \right)^2 dx \tag{3}$$

Feeding the given expression for $T_{(x)}$ in the question

$$\dot{S}_1 = kA \int_0^\delta \left(\frac{-2\beta x}{\alpha - \beta x^2} \right)^2 dx \tag{4}$$

Simplifying

$$\dot{S}_1 = 4kA\beta^2 \int_0^\delta \frac{x^2 \cdot dx}{\left(\alpha - \beta x^2 \right)^2} \tag{5}$$

$$\dot{S}_1 = 4kA\beta^2 \int_0^\delta \frac{dx}{\left(\dfrac{\alpha}{x} - \beta x \right)^2} = 4kA\beta^2 \int_0^\delta \frac{dx}{\left(\beta x - \dfrac{\alpha}{x} \right)^2} \tag{6}$$

$$\dot{S}_1 = \frac{4kA\beta^2}{2\beta} \int_0^\delta \frac{2\beta \cdot dx}{\left(\beta x - \dfrac{\alpha}{x} \right)^2} = 2kA\beta \int_0^\delta \frac{\left[\left(\beta + \dfrac{\alpha}{x^2} \right) + \left(\beta - \dfrac{\alpha}{x^2} \right) \right] \cdot dx}{\left(\beta x - \dfrac{\alpha}{x} \right)^2} \tag{7}$$

$$\dot{S}_1 = 2kA\beta \int_0^\delta \frac{\left(\beta + \dfrac{\alpha}{x^2}\right)\cdot dx}{\left(\beta x - \dfrac{\alpha}{x}\right)^2} + 2kA\beta \int_0^\delta \frac{\left(\beta - \dfrac{\alpha}{x^2}\right)\cdot dx}{\left(\beta x - \dfrac{\alpha}{x}\right)^2}$$

$$\Rightarrow \dot{S}_1 = 2kA\beta \int_0^\delta \frac{\left(\beta + \dfrac{\alpha}{x^2}\right)\cdot dx}{\left(\beta x - \dfrac{\alpha}{x}\right)^2} + 2kA\beta \int_0^\delta \frac{\left(\beta - \dfrac{\alpha}{x^2}\right)\cdot dx}{\beta^2 x^2 + \dfrac{\alpha^2}{x^2} - 2\alpha\beta}$$

$$\Rightarrow \dot{S}_1 = 2kA\beta \int_0^\delta \frac{\left(\beta + \dfrac{\alpha}{x^2}\right)\cdot dx}{\left(\beta x - \dfrac{\alpha}{x}\right)^2} + 2kA\beta \int_0^\delta \frac{\left(\beta - \dfrac{\alpha}{x^2}\right)\cdot dx}{\beta^2 x^2 + \dfrac{\alpha^2}{x^2} - 4\alpha\beta + 2\alpha\beta}$$

$$\Rightarrow \dot{S}_1 = 2kA\beta \int_0^\delta \frac{\left(\beta + \dfrac{\alpha}{x^2}\right)\cdot dx}{\left(\beta x - \dfrac{\alpha}{x}\right)^2} + 2kA\beta \int_0^\delta \frac{\left(\beta - \dfrac{\alpha}{x^2}\right)\cdot dx}{\left(\beta x + \dfrac{\alpha}{x}\right)^2 - \left(2\sqrt{\alpha\beta}\right)^2} \tag{8}$$

Let $\beta x - \dfrac{\alpha}{x} = P$ and $\beta x + \dfrac{\alpha}{x} = V \Rightarrow \left(\beta + \dfrac{\alpha}{x^2}\right)dx = dP$ and $\left(\beta - \dfrac{\alpha}{x^2}\right)dx = dV$

$$\dot{S}_1 = 2kA\beta \int_{-\infty}^{\beta\delta - \frac{\alpha}{\delta}} \frac{dP}{P^2} + 2kA\beta \int_{\infty}^{\beta\delta + \frac{\alpha}{\delta}} \frac{dV}{V^2 - \left(2\sqrt{\alpha\beta}\right)^2} \tag{9}$$

$$\dot{S}_1 = 2kA\beta \left[-\frac{1}{P}\right]_{-\infty}^{\beta\delta - \frac{\alpha}{\delta}} + \frac{2kA\beta}{4\sqrt{\alpha\beta}}\left[\ln\left|\frac{V - 2\sqrt{\alpha\beta}}{V + 2\sqrt{\alpha\beta}}\right|\right]_{\infty}^{\beta\delta + \frac{\alpha}{\delta}} \tag{10}$$

$$\dot{S}_1 = -\frac{2kA\beta}{\beta\delta - \dfrac{\alpha}{\delta}} + \frac{kA}{2}\sqrt{\frac{\beta}{\alpha}}\ln\left|\frac{\beta\delta + \dfrac{\alpha}{\delta} - 2\sqrt{\alpha\beta}}{\beta\delta + \dfrac{\alpha}{\delta} + 2\sqrt{\alpha\beta}}\right| \tag{11}$$

Entropy generation rate associated with internal heat generation across a differential layer is

$$dS_2 = \frac{\dot{q}_g A \cdot dx}{T} \tag{12}$$

The overall entropy generation rate due to internal heat generation will be integral of above

$$\dot{S}_2 = \int d\dot{S}_2 = \int_0^{\delta} \frac{\dot{q}_g A \cdot dx}{T} = \int_0^{\delta} \frac{\dot{q}_g A \cdot dx}{\alpha - \beta x^2} = \dot{q}_g A \int_0^{\delta} \frac{dx}{\alpha - \beta x^2} \tag{13}$$

Simplifying

$$\dot{S}_2 = -\dot{q}_g A \int_0^{\delta} \frac{dx}{\beta x^2 - \alpha} = -\dot{q}_g A \int_0^{\delta} \frac{dx}{\beta \left[x^2 - \left(\sqrt{\frac{\alpha}{\beta}} \right)^2 \right]} \tag{14}$$

$$\dot{S}_2 = -\frac{\dot{q}_g A}{2\sqrt{\frac{\alpha}{\beta}}\beta} \left[\ln \left| \frac{x - \sqrt{\frac{\alpha}{\beta}}}{x + \sqrt{\frac{\alpha}{\beta}}} \right| \right]_0^{\delta} = \frac{\dot{q}_g A}{2\sqrt{\alpha\beta}} \ln \left| \frac{\delta + \sqrt{\frac{\alpha}{\beta}}}{\delta - \sqrt{\frac{\alpha}{\beta}}} \right| \tag{15}$$

The net entropy generation rate will be sum of $\dot{S}_1$ and $\dot{S}_2$

$$\dot{S}_{net} = \dot{S}_1 + \dot{S}_2 = -\frac{2kA\beta}{\beta\delta - \frac{\alpha}{\delta}} + \frac{kA}{2}\sqrt{\frac{\beta}{\alpha}} \ln \left| \frac{\beta\delta + \frac{\alpha}{\delta} - 2\sqrt{\alpha\beta}}{\beta\delta + \frac{\alpha}{\delta} + 2\sqrt{\alpha\beta}} \right| + \frac{\dot{q}_g A}{2\sqrt{\alpha\beta}} \ln \left| \frac{\delta + \sqrt{\frac{\alpha}{\beta}}}{\delta - \sqrt{\frac{\alpha}{\beta}}} \right| \tag{16}$$

Expanding back α, β

$$\dot{S}_{net} = \left[\frac{\dot{q}_g A}{\left(\frac{T_a}{\delta} + \frac{\dot{q}_g}{U} \right)} \right] + \left[\frac{kA}{2} \sqrt{\frac{\frac{\dot{q}_g}{2k}}{T_a + \frac{\dot{q}_g \delta}{U} + \frac{\dot{q}_g \delta^2}{2k}}} \ln \left| \frac{\frac{\dot{q}_g \delta^2}{2k} + T_a + \frac{\dot{q}_g \delta}{U} + \frac{\dot{q}_g \delta^2}{2k} - 2\delta \sqrt{\left(T_a + \frac{\dot{q}_g \delta}{U} + \frac{\dot{q}_g \delta^2}{2k} \right) \frac{\dot{q}_g}{2k}}}{\frac{\dot{q}_g \delta^2}{2k} + T_a + \frac{\dot{q}_g \delta}{U} + \frac{\dot{q}_g \delta^2}{2k} + 2\delta \sqrt{\left(T_a + \frac{\dot{q}_g \delta}{U} + \frac{\dot{q}_g \delta^2}{2k} \right) \frac{\dot{q}_g}{2k}}} \right| \right] +$$

$$\left[\frac{\dot{q}_g A}{2 \sqrt{\left(T_a + \frac{\dot{q}_g \delta}{U} + \frac{\dot{q}_g \delta^2}{2k} \right) \frac{\dot{q}_g}{2k}}} \ln \left| \frac{\delta + \sqrt{\frac{T_a + \frac{\dot{q}_g \delta}{U} + \frac{\dot{q}_g \delta^2}{2k}}{\frac{\dot{q}_g}{2k}}}}{\delta - \sqrt{\frac{T_a + \frac{\dot{q}_g \delta}{U} + \frac{\dot{q}_g \delta^2}{2k}}{\frac{\dot{q}_g}{2k}}}} \right| \right] \tag{17}$$

CASE 2: Entropy generation rate associated with conduction heat transfer across a differential layer is

$$d\dot{S}_1 = -\frac{d\dot{Q}}{T} + \frac{d\dot{Q}}{T + dT} = d\dot{Q} \left[-\frac{1}{T} + \frac{1}{T + dT} \right] = -d\dot{Q} \left[\frac{dT}{T(T + dT)} \right] \tag{18}$$

Since dT is very small, $T + dT \approx T$

$$d\dot{S}_1 = -\left(-k(2\pi rL) \frac{dT}{dr} \right) \left(\frac{dT}{T^2} \right)$$

$$= \left(2\pi rLk \frac{dT}{dr} \right) \left(\frac{dT}{T^2 \, dr} \right) dr = 2\pi rLk \left(\frac{\frac{dT}{dr}}{T} \right)^2 dr \tag{19}$$

The overall entropy generation rate due to conduction will be the integral of above

$$\dot{S}_1 = \int d\dot{S}_1 = \int_0^R 2\pi rLk \left(\frac{\frac{dT}{dr}}{T} \right)^2 dr = 2\pi kL \int_0^R r \left(\frac{\frac{dT}{dr}}{T} \right)^2 dr \tag{20}$$

Feeding the given expression for *T(r)* in the question

$$\dot{S}_1 = 2\pi kL \int_0^R r \left(\frac{-2\beta r}{\alpha - \beta r^2} \right)^2 dr$$

$$\Rightarrow \dot{S}_1 = 8\pi kL\beta^2 \int_0^R \frac{r^3 \cdot dr}{\left(\alpha - \beta r^2 \right)^2} \tag{21}$$

Let $\alpha - \beta r^2 = P \Rightarrow -2\beta r \cdot dr = dP$

$$\dot{S}_1 = 8\pi kL\beta^2 \int_\alpha^{\alpha - \beta R^2} \frac{\left(\dfrac{\alpha - P}{\beta} \right)\left(-\dfrac{dP}{2\beta} \right)}{P^2} \tag{22}$$

Simplifying

$$\dot{S}_1 = 4\pi kL \int_\alpha^{\alpha - \beta R^2} \frac{\left(P - \alpha \right) \cdot dP}{P^2} \tag{23}$$

$$\dot{S}_1 = 4\pi kL \left[\ln|P| + \frac{\alpha}{P} \right]_\alpha^{\alpha - \beta R^2} \tag{24}$$

$$\dot{S}_1 = 4\pi kL \left[\ln\left| \frac{\alpha - \beta R^2}{\alpha} \right| + \frac{\beta R^2}{\alpha - \beta R^2} \right] \tag{25}$$

Entropy generation rate associated with internal heat generation across a differential layer is

$$d\dot{S}_2 = \frac{\dot{q}_g \left(2\pi rL \right) \cdot dr}{T} \tag{26}$$

The overall entropy generation rate due to internal heat generation will be integral of above

$$\dot{S}_2 = \int d\dot{S}_2 = \int_0^R \frac{\dot{q}_g \left(2\pi r L\right)\cdot dr}{\alpha - \beta r^2} \tag{27}$$

Let $r^2 = P \Rightarrow 2r\cdot dr = dP$

$$\dot{S}_2 = \pi\dot{q}_g L \int_0^{R^2} \frac{dP}{\alpha - \beta P} = \pi\dot{q}_g L \left[\frac{\ln\left|\alpha - \beta P\right|}{-\beta}\right]_0^{R^2} = \frac{\pi\dot{q}_g L}{\beta}\ln\left|\frac{\alpha}{\alpha - \beta R^2}\right| \tag{28}$$

The net entropy generation rate will be sum of $\dot{S}_1$ and $\dot{S}_2$

$$\dot{S}_{net} = \dot{S}_1 + \dot{S}_2 = \left[4\pi kL\left\{\ln\left|\frac{\alpha - \beta R^2}{\alpha}\right| + \frac{\beta R^2}{\alpha - \beta R^2}\right\}\right] + \left[\frac{\pi\dot{q}_g L}{\beta}\ln\left|\frac{\alpha}{\alpha - \beta R^2}\right|\right] \tag{29}$$

Expanding back α, β

$$\dot{S}_{net} = \left[4\pi kL\left\{\ln\left|\frac{T_a + \dfrac{\dot{q}_g R}{2U}}{T_a + \dfrac{\dot{q}_g R}{2U} + \dfrac{\dot{q}_g R^2}{4k}}\right| + \frac{\dfrac{\dot{q}_g R^2}{4k}}{T_a + \dfrac{\dot{q}_g R}{2U}}\right\}\right] + \left[4k\pi L\ln\left|\frac{T_a + \dfrac{\dot{q}_g R}{2U} + \dfrac{\dot{q}_g R^2}{4k}}{T_a + \dfrac{\dot{q}_g R}{2U}}\right|\right] \tag{30}$$

CASE 3: Entropy generation rate associated with conduction heat transfer across a differential layer is

$$d\dot{S}_1 = -\frac{d\dot{Q}}{T} + \frac{d\dot{Q}}{T+dT} = d\dot{Q}\left[-\frac{1}{T} + \frac{1}{T+dT}\right] = -d\dot{Q}\left[\frac{dT}{T(T+dT)}\right] \tag{31}$$

Since dT is very small, $T + dT \approx T$

$$\begin{aligned}
d\dot{S}_1 &= -\left[-k\left(4\pi r^2\right)\frac{dT}{dr}\right]\left(\frac{dT}{T^2}\right) \\
&= \left(4\pi k r^2 \frac{dT}{dr}\right)\left(\frac{dT}{T^2 dr}\right)dr \\
&= 4\pi k r^2 \left(\frac{\frac{dT}{dr}}{T}\right)^2 dr
\end{aligned} \tag{32}$$

The overall entropy generation rate due to conduction will be the integral of above

$$\dot{S}_1 = \int d\dot{S}_1 = \int_0^R 4\pi k r^2 \left(\frac{\frac{dT}{dr}}{T}\right)^2 dr = 4\pi k \int_0^R r^2 \left(\frac{\frac{dT}{dr}}{T}\right)^2 dr \tag{33}$$

Feeding the given expression for *T(r)* in the question

$$\dot{S}_1 = 4\pi k \int_0^R r^2 \left(\frac{-2\beta r}{\alpha - \beta r^2}\right)^2 dr$$

$$\Rightarrow \dot{S}_1 = 16\pi k \beta^2 \int_0^R \frac{r^4 \cdot dr}{\left(\alpha - \beta r^2\right)^2} \tag{34}$$

Simplifying

$$\dot{S}_1 = 16\pi k \beta^2 \int_0^R \frac{r^4 \cdot dr}{\alpha^2 + \beta^2 r^4 - 2\alpha\beta r^2}$$

(35)

$$\dot{S}_1 = 16\pi k \int_0^R \frac{\beta^2 r^4 \cdot dr}{\alpha^2 + \beta^2 r^4 - 2\alpha\beta r^2}$$

$$\Rightarrow \dot{S}_1 = 16\pi k \int_0^R \frac{\left(\alpha^2 + \beta^2 r^4 - \alpha^2 + 2\alpha\beta r^2 - 2\alpha\beta r^2\right) \cdot dr}{\alpha^2 + \beta^2 r^4 - 2\alpha\beta r^2}$$

$$\Rightarrow \dot{S}_1 = 16\pi k \int_0^R dr - 16\pi k \alpha^2 \int_0^R \frac{dr}{\alpha^2 + \beta^2 r^4 - 2\alpha\beta r^2} + 32\pi k \alpha\beta \int_0^R \frac{r^2 \cdot dr}{\alpha^2 + \beta^2 r^4 - 2\alpha\beta r^2}$$

$$\Rightarrow \dot{S}_1 = 16\pi k R - 16\pi k \alpha^2 \int_0^R \frac{dr}{\left(\alpha - \beta r^2\right)^2} + 32\pi k \alpha\beta \int_0^R \frac{r^2 \cdot dr}{\left(\alpha - \beta r^2\right)^2}$$

(36)

$$\dot{S}_1 = 16\pi k R - 16\pi k \alpha^2 \int_0^R \frac{dr}{\left(\alpha - \beta r^2\right)^2} - 32\pi k \alpha \int_0^R \frac{-\beta r^2 \cdot dr}{\left(\alpha - \beta r^2\right)^2}$$

$$\Rightarrow \dot{S}_1 = 16\pi k R - 16\pi k \alpha^2 \int_0^R \frac{dr}{\left(\alpha - \beta r^2\right)^2} - 32\pi k \alpha \int_0^R \frac{\left(\alpha - \beta r^2 - \alpha\right) \cdot dr}{\left(\alpha - \beta r^2\right)^2}$$

$$\Rightarrow \dot{S}_1 = 16\pi k R - 16\pi k \alpha^2 \int_0^R \frac{dr}{\left(\alpha - \beta r^2\right)^2} - 32\pi k \alpha \int_0^R \frac{dr}{\alpha - \beta r^2} + 32\pi k \alpha^2 \int_0^R \frac{dr}{\left(\alpha - \beta r^2\right)^2}$$

$$\Rightarrow \dot{S}_1 = 16\pi k R + 16\pi k \alpha^2 \int_0^R \frac{dr}{\left(\alpha - \beta r^2\right)^2} + \frac{32\pi k \alpha}{\beta} \int_0^R \frac{dr}{r^2 - \left(\sqrt{\dfrac{\alpha}{\beta}}\right)^2}$$

(37)

$$\dot{S}_1 = 16\pi kR + 16\pi k\alpha^2 \int_0^R \frac{dr}{\left(\alpha - \beta r^2\right)^2} + \frac{32\pi k\alpha}{2\beta\sqrt{\dfrac{\alpha}{\beta}}} \ln\left|\frac{R - \sqrt{\dfrac{\alpha}{\beta}}}{R + \sqrt{\dfrac{\alpha}{\beta}}}\right|$$

$$\Rightarrow \dot{S}_1 = 16\pi kR + 16\pi k\alpha^2 \int_0^R \frac{dr}{\left(\alpha - \beta r^2\right)^2} + 16\pi k\sqrt{\dfrac{\alpha}{\beta}} \ln\left|\frac{R - \sqrt{\dfrac{\alpha}{\beta}}}{R + \sqrt{\dfrac{\alpha}{\beta}}}\right| \tag{38}$$

Let the indefinite integral $\dfrac{dr}{\left(\alpha - \beta r^2\right)^2}$ be termed "Integral". It can be evaluated as shown below

$$\text{Integral} = \int \frac{dr}{\left(\alpha - \beta r^2\right)^2}$$

$$\Rightarrow \text{Integral} = \int \frac{\dfrac{dr}{r^2}}{\left(\dfrac{\alpha}{r} - \beta r\right)^2} = \int \frac{\dfrac{dr}{r^2}}{\left(\beta r - \dfrac{\alpha}{r}\right)^2}$$

$$\Rightarrow \text{Integral} = \frac{1}{2\alpha} \int \frac{2\alpha \dfrac{dr}{r^2}}{\left(\beta r - \dfrac{\alpha}{r}\right)^2}$$

$$\Rightarrow \text{Integral} = \frac{1}{2\alpha} \int \frac{\left[\left(\beta + \dfrac{\alpha}{r^2}\right) - \left(\beta - \dfrac{\alpha}{r^2}\right)\right]dr}{\left(\beta r - \dfrac{\alpha}{r}\right)^2}$$

$$\Rightarrow \text{Integral} = \frac{1}{2\alpha} \int \frac{\left(\beta + \dfrac{\alpha}{r^2}\right)dr}{\left(\beta r - \dfrac{\alpha}{r}\right)^2} - \frac{1}{2\alpha} \int \frac{\left(\beta - \dfrac{\alpha}{r^2}\right)dr}{\left(\beta r - \dfrac{\alpha}{r}\right)^2} \tag{39}$$

$$\text{Integral} = \frac{1}{2\alpha}\int \frac{\left(\beta + \dfrac{\alpha}{r^2}\right)dr}{\left(\beta r - \dfrac{\alpha}{r}\right)^2} - \frac{1}{2\alpha}\int \frac{\left(\beta - \dfrac{\alpha}{r^2}\right)dr}{\beta^2 r^2 + \dfrac{\alpha^2}{r^2} - 2\alpha\beta}$$

$$\Rightarrow \text{Integral} = \frac{1}{2\alpha}\int \frac{\left(\beta + \dfrac{\alpha}{r^2}\right)dr}{\left(\beta r - \dfrac{\alpha}{r}\right)^2} - \frac{1}{2\alpha}\int \frac{\left(\beta - \dfrac{\alpha}{r^2}\right)dr}{\beta^2 r^2 + \dfrac{\alpha^2}{r^2} + 2\alpha\beta - 4\alpha\beta}$$

$$\Rightarrow \text{Integral} = \frac{1}{2\alpha}\int \frac{\left(\beta + \dfrac{\alpha}{r^2}\right)dr}{\left(\beta r - \dfrac{\alpha}{r}\right)^2} - \frac{1}{2\alpha}\int \frac{\left(\beta - \dfrac{\alpha}{r^2}\right)dr}{\left(\beta r + \dfrac{\alpha}{r}\right)^2 - \left(2\sqrt{\alpha\beta}\right)^2} \tag{40}$$

Let $\beta r - \dfrac{\alpha}{r} = P$ and $\beta r + \dfrac{\alpha}{r} = V \Rightarrow \left(\beta + \dfrac{\alpha}{r^2}\right)dr = dP$ and $\left(\beta - \dfrac{\alpha}{r^2}\right)dr = dV$

$$\text{Integral} = \frac{1}{2\alpha}\int \frac{dP}{P^2} - \frac{1}{2\alpha}\int \frac{dV}{V^2 - \left(2\sqrt{\alpha\beta}\right)^2}$$

$$\Rightarrow \text{Integral} = -\frac{1}{2\alpha P} - \frac{1}{4\alpha\sqrt{\alpha\beta}}\ln\left|\frac{V - 2\sqrt{\alpha\beta}}{V + 2\sqrt{\alpha\beta}}\right|$$

$$\Rightarrow \text{Integral} = \frac{r}{2\alpha\left(\alpha - \beta r^2\right)} + \frac{1}{4\alpha\sqrt{\alpha\beta}}\ln\left|\frac{\beta r^2 + \alpha + 2r\sqrt{\alpha\beta}}{\beta r^2 + \alpha - 2r\sqrt{\alpha\beta}}\right| \tag{41}$$

Feeding equation 41 in 38

$$\dot{S}_1 = 16\pi k R + 16\pi k \alpha^2 \left[\frac{R}{2\alpha\left(\alpha - \beta R^2\right)} + \frac{1}{4\alpha\sqrt{\alpha\beta}}\ln\left|\frac{\beta R^2 + \alpha + 2R\sqrt{\alpha\beta}}{\beta R^2 + \alpha - 2R\sqrt{\alpha\beta}}\right|\right] +$$

$$16\pi k \sqrt{\frac{\alpha}{\beta}}\ln\left|\frac{R - \sqrt{\dfrac{\alpha}{\beta}}}{R + \sqrt{\dfrac{\alpha}{\beta}}}\right| \tag{42}$$

Entropy generation rate associated with internal heat generation across a differential layer is

$$dS_2 = \frac{\dot{q}_g\left(4\pi r^2\right)\cdot dr}{T} = \frac{4\pi\dot{q}_g r^2 \cdot dr}{\alpha - \beta r^2} \tag{43}$$

The overall entropy generation rate due to internal heat generation will be integral of above

$$\dot{S}_2 = \int d\dot{S}_2 = \int_0^R \frac{4\pi\dot{q}_g r^2 \cdot dr}{\alpha - \beta r^2} = 4\pi\dot{q}_g\int_0^R \frac{r^2 \cdot dr}{\alpha - \beta r^2} \tag{44}$$

Simplifying

$$\dot{S}_2 = \frac{4\pi\dot{q}_g}{-\beta}\int_0^R \frac{-\beta r^2 \cdot dr}{\alpha - \beta r^2}$$

$$\Rightarrow \dot{S}_2 = \frac{4\pi\dot{q}_g}{-\beta}\int_0^R \frac{\left(\alpha - \beta r^2 - \alpha\right)\cdot dr}{\alpha - \beta r^2}$$

$$\Rightarrow \dot{S}_2 = \frac{4\pi\dot{q}_g\alpha}{\beta}\int_0^R \frac{dr}{\alpha - \beta r^2} - \frac{4\pi\dot{q}_g}{\beta}\int_0^R dr \tag{45}$$

$$\dot{S}_2 = -\frac{4\pi\dot{q}_g\alpha}{\beta^2}\int_0^R \frac{dr}{r^2 - \left(\sqrt{\dfrac{\alpha}{\beta}}\right)^2} - \frac{4\pi\dot{q}_g R}{\beta}$$

$$\Rightarrow \dot{S}_2 = -\frac{4\pi\dot{q}_g\alpha}{2\beta^2\sqrt{\dfrac{\alpha}{\beta}}}\ln\left|\frac{R-\sqrt{\dfrac{\alpha}{\beta}}}{R+\sqrt{\dfrac{\alpha}{\beta}}}\right| - \frac{4\pi\dot{q}_g R}{\beta}$$

$$\Rightarrow \dot{S}_2 = \frac{2\pi\dot{q}_g\sqrt{\alpha}}{\sqrt{\beta^3}}\ln\left|\frac{R+\sqrt{\dfrac{\alpha}{\beta}}}{R-\sqrt{\dfrac{\alpha}{\beta}}}\right| - \frac{4\pi\dot{q}_g R}{\beta} \tag{46}$$

The net entropy generation rate will be sum of $\dot{S}_1$ and $\dot{S}_2$

$$\dot{S}_{net} = \dot{S}_1 + \dot{S}_2 = \left[16\pi kR\right] + \left[16\pi k\alpha^2\left\{\frac{R}{2\alpha(\alpha-\beta R^2)} + \frac{1}{4\alpha\sqrt{\alpha\beta}}\ln\left|\frac{\beta R^2 + \alpha + 2R\sqrt{\alpha\beta}}{\beta R^2 + \alpha - 2R\sqrt{\alpha\beta}}\right|\right\}\right] +$$

$$\left[16\pi k\sqrt{\frac{\alpha}{\beta}}\ln\left|\frac{R-\sqrt{\dfrac{\alpha}{\beta}}}{R+\sqrt{\dfrac{\alpha}{\beta}}}\right|\right] + \left[\frac{2\pi\dot{q}_g\sqrt{\alpha}}{\sqrt{\beta^3}}\ln\left|\frac{R+\sqrt{\dfrac{\alpha}{\beta}}}{R-\sqrt{\dfrac{\alpha}{\beta}}}\right|\right] - \left[\frac{4\pi\dot{q}_g R}{\beta}\right] \tag{47}$$

Expanding back α, β

$$
\dot{S}_{net} = \left[16\pi k \left(T_a + \frac{\dot{q}_g R}{3U} + \frac{\dot{q}_g R^2}{6k}\right)^2 \left\{ \frac{R}{2\left(T_a + \frac{\dot{q}_g R}{3U} + \frac{\dot{q}_g R^2}{6k}\right)\left(T_a + \frac{\dot{q}_g R}{3U}\right)} + \frac{1}{4\left(T_a + \frac{\dot{q}_g R}{3U} + \frac{\dot{q}_g R^2}{6k}\right)^{\frac{3}{2}}\sqrt{\frac{\dot{q}_g}{6k}}} \ln \frac{\frac{\dot{q}_g R^2}{3k} + T_a + \frac{\dot{q}_g R}{3U} + 2R\sqrt{\left(T_a + \frac{\dot{q}_g R}{3U} + \frac{\dot{q}_g R^2}{6k}\right)\frac{\dot{q}_g}{6k}}}{\frac{\dot{q}_g R^2}{3k} + T_a + \frac{\dot{q}_g R}{3U} - 2R\sqrt{\left(T_a + \frac{\dot{q}_g R}{3U} + \frac{\dot{q}_g R^2}{6k}\right)\frac{\dot{q}_g}{6k}}} \right\} \right] +
$$

$$
\left[16\pi k \sqrt{\frac{T_a + \frac{\dot{q}_g R}{3U} + \frac{\dot{q}_g R^2}{6k}}{\frac{\dot{q}_g}{6k}}} \; \ln \frac{R - \sqrt{\frac{T_a + \frac{\dot{q}_g R}{3U} + \frac{\dot{q}_g R^2}{6k}}{\frac{\dot{q}_g}{6k}}}}{R + \sqrt{\frac{T_a + \frac{\dot{q}_g R}{3U} + \frac{\dot{q}_g R^2}{6k}}{\frac{\dot{q}_g}{6k}}}} \right] +
$$

$$
\left[\frac{2\pi \dot{q}_g \sqrt{T_a + \frac{\dot{q}_g R}{3U} + \frac{\dot{q}_g R^2}{6k}}}{\left(\frac{\dot{q}_g}{6k}\right)^{\frac{3}{2}}} \; \ln \frac{R + \sqrt{\frac{T_a + \frac{\dot{q}_g R}{3U} + \frac{\dot{q}_g R^2}{6k}}{\frac{\dot{q}_g}{6k}}}}{R - \sqrt{\frac{T_a + \frac{\dot{q}_g R}{3U} + \frac{\dot{q}_g R^2}{6k}}{\frac{\dot{q}_g}{6k}}}} \right] - [8\pi k R]
$$

$$(48)$$

Question 17

A cuboidal rod of cross-sectional area A, width b, length L has one end of it immersed in a large thermal reservoir at temperature T_0. The other end is insulated. The sides and top surface are also insulated, while the base isn't. Below the rod, a duct is installed in which air flows steadily at a mass flow rate $\dot{m}_a$. Air enters at ambient temperature T_a as shown. The duct is insulated at base and sides. Find the outlet temperature of air under steady state. The specific heat capacity of air is c_{pa} and overall heat transfer coefficient from rod to air is U. The thermal conductivity of the rod material is k. Neglect heat conduction, viscous effects and pressure gradients in air stream.

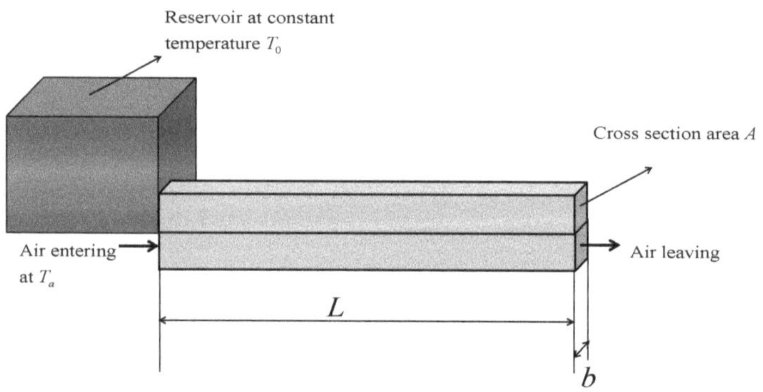

Solution

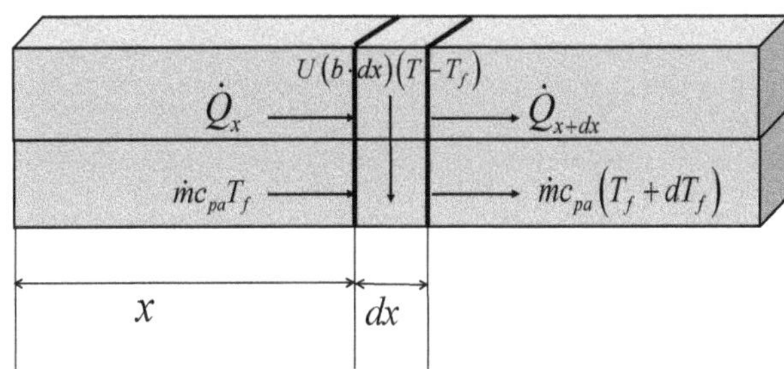

Applying law of energy conservation for the differential elements of rod and air stream shown

$$\dot{Q}_x - \dot{Q}_{x+dx} - U(b \cdot dx)(T - T_f) = 0$$

$$\Rightarrow \dot{Q}_x - \left(\dot{Q}_x + d\dot{Q}_x\right) - U(b \cdot dx)(T - T_f) = 0$$

$$\Rightarrow -d\dot{Q}_x - U(b \cdot dx)(T - T_f) = 0$$

$$\Rightarrow -d\left(-kA\frac{dT}{dx}\right) - U(b \cdot dx)(T - T_f) = 0$$

$$\Rightarrow -\frac{d}{dx}\left(-kA\frac{dT}{dx}\right) - Ub(T - T_f) = 0$$

$$\Rightarrow \frac{d^2T}{dx^2} - a_0(T - T_f) = 0 \tag{1}$$

$$U(b \cdot dx)(T - T_f) + \dot{m}_a c_{pa} T_f - \left(\dot{m}_a c_{pa} T_f + \dot{m}_a c_{pa} dT_f\right) = 0$$

$$\Rightarrow \frac{dT_f}{dx} = b_0(T - T_f) \tag{2}$$

Where $a_0 = \dfrac{Ub}{kA}, b_0 = \dfrac{Ub}{\dot{m}_a c_{pa}}$. Feeding T from equation 2 in 1

$$\frac{d^2}{dx^2}\left(T_f + \frac{1}{b_0}\frac{dT_f}{dx}\right) - a_0\left(\frac{1}{b_0}\frac{dT_f}{dx}\right) = 0$$

$$\Rightarrow \frac{d^2 T_f}{dx^2} + \frac{1}{b_0}\frac{d^3 T_f}{dx^3} - \frac{a_0}{b_0}\frac{dT_f}{dx} = 0$$

$$\Rightarrow \frac{d^3 T_f}{dx^3} + b_0\frac{d^2 T_f}{dx^2} - a_0\frac{dT_f}{dx} = 0 \tag{3}$$

Separating the variables and integrating both sides

$$\frac{d}{dx}\left(\frac{d^2 T_f}{dx^2}\right) + b_0\frac{d}{dx}\left(\frac{dT_f}{dx}\right) - a_0\frac{dT_f}{dx} = 0$$

$$\Rightarrow d\left(\frac{d^2 T_f}{dx^2}\right) + b_0 d\left(\frac{dT_f}{dx}\right) - a_0 dT_f = 0$$

$$\Rightarrow \int d\left(\frac{d^2 T_f}{dx^2}\right) + b_0\int d\left(\frac{dT_f}{dx}\right) - a_0\int dT_f = C_0$$

$$\Rightarrow \frac{d^2 T_f}{dx^2} + b_0\frac{dT_f}{dx} - a_0 T_f = C_0 \tag{4}$$

Where C_0 is an arbitrary integration constant. Equation 4 is an inhomogenous second order linear ordinary differential equation with constant coefficients. Its general solution can be given by the method of variation of parameters as shown below.

$$T_f = C_1 e^{s_1 x} + C_2 e^{s_2 x} + e^{s_2 x}\int \frac{e^{s_1 x}\cdot C_0\cdot dx}{\left(s_2 - s_1\right)e^{(s_1+s_2)x}} - e^{s_1 x}\int \frac{e^{s_2 x}\cdot C_0\cdot dx}{\left(s_2 - s_1\right)e^{(s_1+s_2)x}} \tag{5}$$

Where C_1, C_2 are integration constants and $s_1, s_2 = \dfrac{-b_0 \pm \sqrt{b_0^2 + 4a_0}}{2}$.

Simplifying equation 5,

$$T_f = C_1 e^{s_1 x} + C_2 e^{s_2 x} + \frac{C_0}{(s_2 - s_1)} e^{s_2 x} \int e^{-s_2 x} \cdot dx - \frac{C_0}{(s_2 - s_1)} e^{s_1 x} \int e^{-s_1 x} \cdot dx$$

$$\Rightarrow T_f = C_1 e^{s_1 x} + C_2 e^{s_2 x} + \frac{C_0}{(s_2 - s_1)} e^{s_2 x} \left(\frac{e^{-s_2 x}}{-s_2} \right) - \frac{C_0}{(s_2 - s_1)} e^{s_1 x} \left(\frac{e^{-s_1 x}}{-s_1} \right)$$

$$\Rightarrow T_f = C_1 e^{s_1 x} + C_2 e^{s_2 x} + \frac{C_0}{(s_2 - s_1)} \left(\frac{1}{s_1} - \frac{1}{s_2} \right)$$

$$\Rightarrow T_f = C_1 e^{s_1 x} + C_2 e^{s_2 x} + \frac{C_0}{s_1 s_2} \tag{6}$$

Feeding equation 6 in 2

$$T = C_1 e^{s_1 x} + C_2 e^{s_2 x} + \frac{C_0}{s_1 s_2} + \frac{\left(C_1 s_1 e^{s_1 x} + C_2 s_2 e^{s_2 x} \right)}{b_0} \tag{7}$$

It can be seen that $b_0 = -(s_1 + s_2)$. Therefore, equation 7 can be better written as

$$T = C_1 e^{s_1 x} + C_2 e^{s_2 x} + \frac{C_0}{s_1 s_2} - \frac{\left(C_1 s_1 e^{s_1 x} + C_2 s_2 e^{s_2 x} \right)}{s_1 + s_2}$$

$$\Rightarrow T = \frac{C_1 s_2 e^{s_1 x}}{s_1 + s_2} + \frac{C_2 s_1 e^{s_2 x}}{s_1 + s_2} + \frac{C_0}{s_1 s_2} \tag{8}$$

Having evaluated the general solutions for the temperature distributions *T(x)* and *Tf(x)*, the next step is to find out the unknown integration constants. The pertinent boundary conditions for this are:

$$\text{B.C 1: } T(0) = T_0$$

$$\text{B.C 2: } \frac{dT}{dx}(L) = 0$$

$$\text{B.C 3: } T_f(0) = T_a \tag{9}$$

Substituting equations 6 and 8 in 9

$$\frac{C_1 s_2}{s_1 + s_2} + \frac{C_2 s_1}{s_1 + s_2} + \frac{C_0}{s_1 s_2} = T_0$$

(10)

$$\frac{C_1 s_1 s_2 e^{s_1 L}}{s_1 + s_2} + \frac{C_2 s_1 s_2 e^{s_2 L}}{s_1 + s_2} = 0$$
$$\Rightarrow C_1 e^{s_1 L} + C_2 e^{s_2 L} = 0$$

(11)

$$C_1 + C_2 + \frac{C_0}{s_1 s_2} = T_a$$

(12)

Feeding C_2 from equation 11 in 10 and 12

$$\frac{C_1 s_2}{s_1 + s_2} - C_1 \frac{e^{s_1 L}}{e^{s_2 L}} \frac{s_1}{s_1 + s_2} + \frac{C_0}{s_1 s_2} = T_0$$

(13)

$$C_1 - C_1 \frac{e^{s_1 L}}{e^{s_2 L}} + \frac{C_0}{s_1 s_2} = T_a$$

(14)

Subtracting equation 14 from 13

$$\frac{C_1}{s_1 + s_2}\left(s_2 e^{(s_1 - s_2)L} - s_1\right) = T_0 - T_a$$
$$\Rightarrow C_1 = \frac{(T_0 - T_a)(s_1 + s_2)}{s_2 e^{(s_1 - s_2)L} - s_1}$$

(15)

Feeding equation 15 in 11

$$C_2 = -\frac{(T_0 - T_a)(s_1 + s_2)e^{(s_1 - s_2)L}}{s_2 e^{(s_1 - s_2)L} - s_1}$$

(16)

Feeding equations 15 and 16 in 12

$$C_0 = T_a s_1 s_2 - \frac{\left(T_0 - T_a\right)\left(s_1 + s_2\right)s_1 s_2 \left(1 - e^{(s_1 - s_2)L}\right)}{s_2 e^{(s_1 - s_2)L} - s_1} \tag{17}$$

Feeding equations 15, 16, 17 in 6 and taking $x=L$ in the expression obtained

$$\left(T_f\right)_{x=L} = T_a - \frac{\left(T_0 - T_a\right)\left(s_1 + s_2\right)\left(1 - e^{(s_1 - s_2)L}\right)}{s_2 e^{(s_1 - s_2)L} - s_1} \tag{18}$$

Expanding back s_1, s_2

$$\left(T_f\right)_{x=L} = T_a - \frac{2\left(T_0 - T_a\right)b_0 \left(1 - e^{L\sqrt{b_0^2 + 4a_0}}\right)}{e^{L\sqrt{b_0^2 + 4a_0}}\sqrt{b_0^2 + 4a_0} - b_0 + b_0 e^{L\sqrt{b_0^2 + 4a_0}} + \sqrt{b_0^2 + 4a_0}} \tag{19}$$

Expanding back a_0, b_0

$$\left(T_f\right)_{x=L} = \left[T_a\right] - \frac{\dfrac{2Ub\left(T_0 - T_a\right)}{\dot{m}_a c_{pa}}\left(1 - e^{L\sqrt{\left(\frac{Ub}{\dot{m}_a c_{pa}}\right)^2 + \frac{4Ub}{kA}}}\right)}{\left\{e^{L\sqrt{\left(\frac{Ub}{\dot{m}_a c_{pa}}\right)^2 + \frac{4Ub}{kA}}}\sqrt{\left(\frac{Ub}{\dot{m}_a c_{pa}}\right)^2 + \frac{4Ub}{kA}}\right\} - \left\{\dfrac{Ub}{\dot{m}_a c_{pa}}\right\} + \left\{\dfrac{Ube^{L\sqrt{\left(\frac{Ub}{\dot{m}_a c_{pa}}\right)^2 + \frac{4Ub}{kA}}}}{\dot{m}_a c_{pa}}\right\} + \left\{\sqrt{\left(\frac{Ub}{\dot{m}_a c_{pa}}\right)^2 + \frac{4Ub}{kA}}\right\}} \tag{20}$$

Question 18

A solid rod, fully insulated, is initially at temperature T_i. At $t=0$, one end's insulation is removed, and a box at temperature T_0 ($<T_i$) is placed in contact with it. The box is very small such that temperature gradients inside it can be ignored. Further, the box receives heat from the rod alone and all other surfaces of it are insulated. Find temperature of box as a function of time.

Rod length, mass density, thermal conductivity, specific heat capacity and cross section area are L, ρ, k, c_p, A

Box's mass and specific heat capacity are m, c_{pb}

Solution

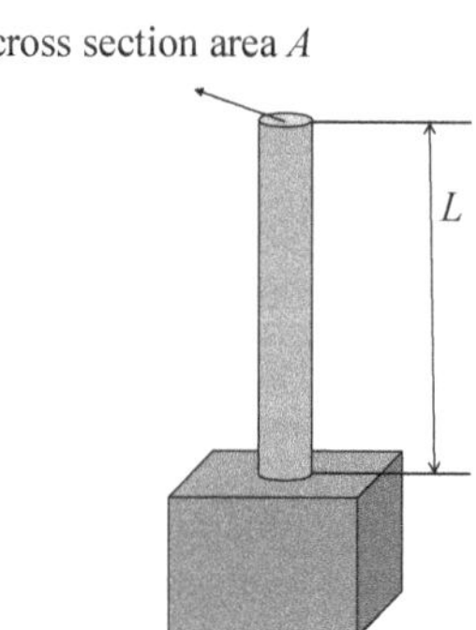

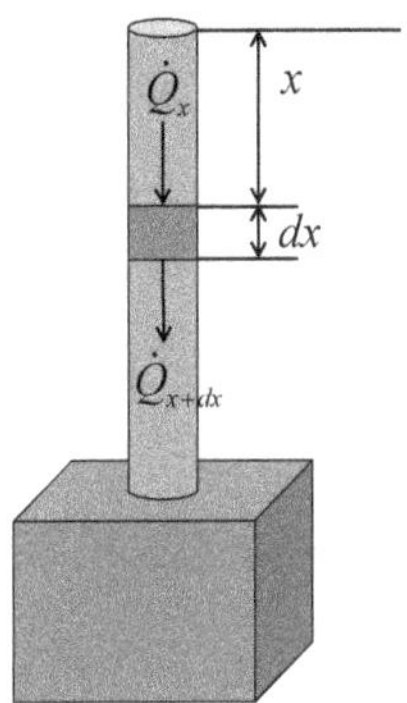

Applying law of energy conservation for the differential element shown

$$\dot{Q}_x - \dot{Q}_{x+dx} = \frac{\partial}{\partial t}\left(E_{int\,ernal}\right)$$

$$\Rightarrow \dot{Q}_x - \left(\dot{Q}_x + \frac{\partial \dot{Q}_x}{\partial x}dx\right) = \frac{\partial}{\partial t}\left(E_{int\,ernal}\right)$$

$$\Rightarrow -\frac{\partial}{\partial x}\left(-kA\frac{\partial T}{\partial x}\right)dx = \frac{\partial}{\partial t}\left(\rho A \cdot dx \cdot c_p T\right)$$

$$\Rightarrow \frac{\partial^2 T}{\partial x^2} = \frac{\rho c_p}{k}\frac{\partial T}{\partial t} \tag{1}$$

The relevant boundary and initial conditions are

$$\text{B.C 1: } \frac{\partial T}{\partial x}(0,t) = 0$$

$$\text{B.C 2: } -kA\frac{\partial T}{\partial x}(L,t) = mc_{pb}\frac{\partial T}{\partial t}(L,t)$$

$$\text{I.C : } T(x,0) = T_i \tag{2}$$

Assuming a product solution of the form *T=F(x)G(t)* and feeding it in equation 1

$$F''G = aFG'$$

$$\Rightarrow \frac{F''}{F} = \frac{aG'}{G} \tag{3}$$

Where $a = \dfrac{\rho c_p}{k}$ Since left hand side is a function of x alone and right-hand side of t alone, they can be equal to each other only if each is equal to a constant. That constant can be either positive, zero or negative. It can be seen that if this constant is 0, then the function G will be a constant. This will mean that temperature in the rod is independent of time which is not physically meaningful. If that constant is positive, then G shall be an increasing function of time. This is again not physically meaningful because after a certain time, steady state will be attained and so the temporal part of the solution must decrease with time. Therefore, the only possibility consistent with the physical nature of the problem is for that constant to be negative. Let this constant be $-p^2$

$$\frac{F''}{F} = \frac{aG'}{G} = -p^2 \tag{4}$$

Equation 4 can be decomposed into two ordinary differential equations

$$F'' + p^2 F = 0 \text{ and}$$

$$\frac{G'}{G} = -\frac{p^2}{a} \tag{5}$$

Both of these are standard ordinary differential equations whose general solutions are well known. They are

$$F = C_1 \cos px + C_2 \sin px \text{ and}$$

$$G = C_3 e^{-\frac{p^2 t}{a}} \tag{6}$$

The product solution T can thus be written as

$$T = \left(C_1 \cos px + C_2 \sin px\right)C_3 e^{-\frac{p^2 t}{a}} \tag{7}$$

We have two boundary conditions and one initial condition. And the number of unknown constants seems to be 4 : C_1, C_2, C_3, p. It appears that available data is insufficient. However, careful observation tells that the number of independent constants is 3: C_{11}, C_{22}, p. Hence, equation 7 can be expressed as

$$T = \left(C_{11} \cos px + C_{22} \sin px\right)e^{-\frac{p^2 t}{a}} \tag{8}$$

Where $C_{11} = C_1 C_3, C_{22} = C_2 C_3$. Applying 1st boundary condition

$$C_{22} p e^{-\frac{p^2 t}{a}} = 0 \text{ for all } t$$
$$\Rightarrow C_{22} = 0 \tag{9}$$

Applying 2nd boundary condition

$$-kA\left(-pC_{11} \sin\left(pL\right)e^{-\frac{p^2 t}{a}}\right) = mc_{pb}\left(-\frac{p^2}{a}C_{11}\cos\left(pL\right)e^{-\frac{p^2 t}{a}}\right)$$

$$\Rightarrow \tan\left(pL\right) = -p\left(\frac{mc_{pb}}{kAa}\right) \tag{10}$$

There are infinitely many solutions to equation 10. Therefore, the general solution *T(x,t)* will be sum of all possible solutions, each corresponding to one value of p

$$T = \sum_{n=1}^{\infty} C_{11-n} \cos\left(p_n x\right) e^{-\frac{p_n^2 t}{a}}$$

(11)

Applying initial condition

$$T_i = \sum_{n=1}^{\infty} C_{11-n} \cos\left(p_n x\right)$$

(12)

The left-hand side of equation 12 can be expressed as a generalized fourier cosine series

$$\sum_{n=1}^{\infty} H_n \cos\left(p_n x\right) = \sum_{n=1}^{\infty} C_{11-n} \cos\left(p_n x\right)$$

$$where \; H_n = \frac{\displaystyle\int_0^L T_i \cos\left(p_n x\right) \cdot dx}{\displaystyle\int_0^L \cos^2\left(p_n x\right) \cdot dx} = \frac{\displaystyle\int_0^L T_i \cos\left(p_n x\right) \cdot dx}{\displaystyle\int_0^L \left\{\frac{1 + \cos\left(2 p_n x\right)}{2}\right\} \cdot dx}$$

$$= \cdot \frac{T_i \left[\dfrac{\sin\left(p_n x\right)}{p_n}\right]_0^L}{\left[\dfrac{x}{2} + \dfrac{\sin\left(2 p_n x\right)}{4 p_n}\right]_0^L} = \frac{\dfrac{T_i \sin\left(p_n L\right)}{p_n}}{\dfrac{L}{2} + \dfrac{\sin\left(2 p_n L\right)}{4 p_n}}$$

$$= \frac{4 T_i \sin\left(p_n L\right)}{2 p_n L + \sin\left(2 p_n L\right)}$$

(13)

Equation 13 can be expressed as

$$\sum_{n=1}^{\infty} H_n \cos(p_n x) = \sum_{n=1}^{\infty} C_{11-n} \cos(p_n x)$$

$$\Rightarrow \sum_{n=1}^{\infty} H_n \cos(p_n x) - \sum_{n=1}^{\infty} C_{11-n} \cos(p_n x) = 0$$

$$\Rightarrow \sum_{n=1}^{\infty} (H_n - C_{11-n}) \cos(p_n x) = 0 \tag{14}$$

Since LHS=0 for all x, therefore

$$C_{11-n} = H_n = \frac{4T_i \sin(p_n L)}{2 p_n L + \sin(2 p_n L)} \tag{15}$$

Feeding equation 15 in 11

$$T = \sum_{n=1}^{\infty} \frac{4T_i \sin(p_n L) \cos(p_n x) e^{-\frac{p_n^2 t}{a}}}{2 p_n L + \sin(2 p_n L)} \tag{16}$$

The temperature of box is the value of T at $x=L$

Box temperature at time

$$t = \sum_{n=1}^{\infty} \frac{4T_i \sin(p_n L) \cos(p_n L) e^{-\frac{p_n^2 t}{a}}}{2 p_n L + \sin(2 p_n L)} = 2T_i \sum_{n=1}^{\infty} \frac{\sin(2 p_n L) e^{-\frac{p_n^2 t}{a}}}{2 p_n L + \sin(2 p_n L)}$$

where roots of the transcedental equation

$$\tan(pL) = -p \left(\frac{mc_{pb}}{kAa} \right) \text{ gives values of } p_n \ (n = 1, 2, 3.....) \tag{17}$$

Expanding back, *a*

$$\text{Box temperature at time } t = 2T_i \sum_{n=1}^{\infty} \frac{\sin\left(2p_n L\right) e^{-\frac{kp_n^2 t}{\rho c_p}}}{2p_n L + \sin\left(2p_n L\right)}$$

where roots of the transcedental equation

$$\tan\left(pL\right) = -p\left(\frac{mc_{pb}}{A\rho c_p}\right) \text{ gives values of } p_n \left(n = 1, 2, 3.....\right) \tag{18}$$

Question 19

A tank filled with sand is kept in open sun as shown. It is insulated at the base and at the walls but open at the top. Find temperature of sand as a function of space and time if initially, the entire sand bed is at temperature T_0. The surface loses heat to surroundings via an overall heat loss coefficient as $\dot{Q}_{loss} = h_{sur} A \left(T_{surface} - T_a \right)$. Thermal conductivity, mass density and specific heat capacity of sand are k, ρ, c_p. The height of the tank is H. The solar radiative intensity and ambient temperature for therelevanttime period are given by the functions: $I = \sum_{i=1}^{5} \alpha_i e^{\beta_i t}, T_a = \sum_{i=1}^{5} \gamma_i e^{\mu_i t}$. Consider surrounding temperature T_a to remain constant because of large size of surroundings.

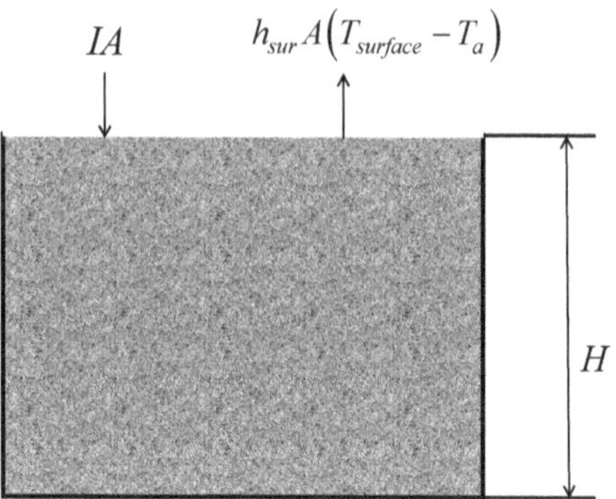

Solution

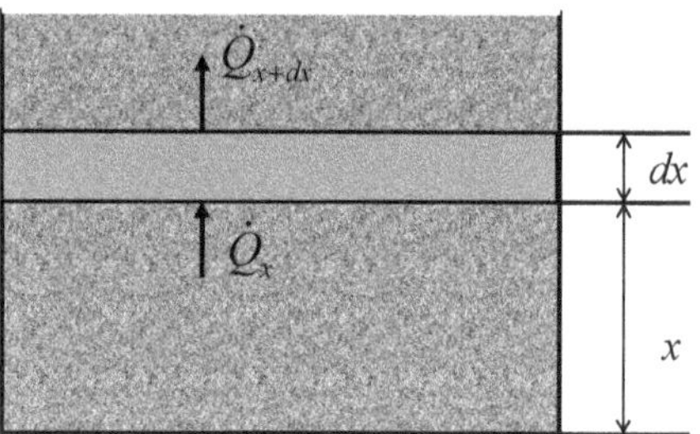

Applying law of energy conservation for the differential element shown

$$\dot{Q}_x - \dot{Q}_{x+dx} = \frac{\partial}{\partial t}\left(E_{internal}\right)$$

$$\Rightarrow \dot{Q}_x - \left(\dot{Q}_x + \frac{\partial \dot{Q}_x}{\partial x} dx\right) = \frac{\partial}{\partial t}\left(E_{internal}\right)$$

$$\Rightarrow -\frac{\partial}{\partial x}\left(-kA\frac{\partial T}{\partial x}\right)dx = \frac{\partial}{\partial t}\left(\rho A \cdot dx \cdot c_p T\right)$$

$$\Rightarrow \frac{\partial^2 T}{\partial x^2} = \frac{\rho c_p}{k}\frac{\partial T}{\partial t} \tag{1}$$

The relevant boundary and initial conditions are:

$$B.C \ 1: \ \frac{\partial T}{\partial x}(0,t) = 0$$

$$B.C \ 2: IA - kA\frac{\partial T}{\partial x}(H,t) = h_{sur}A\left[T(H,t) - T_a(t)\right]$$

$$\Rightarrow \frac{\sum_{i=1}^{5}\alpha_i e^{\beta_i t}}{h_{sur}} - \frac{k}{h_{sur}}\frac{\partial T}{\partial x}(H,t) = T(H,t) - \sum_{i=1}^{5}\gamma_i e^{\mu_i t}$$

$$I.C \ : T(x,0) = T_0 \tag{2}$$

Introducing the change of variable $V = T - \sum_{i=1}^{5}\gamma_i e^{\mu_i t} - \dfrac{\sum_{i=1}^{5}\alpha_i e^{\beta_i t}}{h_{sur}}$, the problem transforms to

$$\frac{\partial^2 V}{\partial x^2} = \frac{\rho c_p}{k}\left[\frac{\partial V}{\partial t} + \sum_{i=1}^{5}\gamma_i\mu_i e^{\mu_i t} + \frac{\sum_{i=1}^{5}\alpha_i\beta_i e^{\beta_i t}}{h_{sur}}\right]$$

$$\Rightarrow \frac{\partial^2 V}{\partial x^2} - \frac{\rho c_p}{k}\left[\sum_{i=1}^{5}\gamma_i\mu_i e^{\mu_i t} + \frac{\sum_{i=1}^{5}\alpha_i\beta_i e^{\beta_i t}}{h_{sur}}\right] = \frac{\rho c_p}{k}\frac{\partial V}{\partial t} \tag{3}$$

$$B.C \ 1: \ \frac{\partial V}{\partial x}(0,t) = 0$$

$$B.C \ 2: -\frac{k}{h_{sur}}\frac{\partial V}{\partial x}(H,t) = V(H,t)$$

$$I.C \ : V(x,0) = T_0 - \sum_{i=1}^{5}\gamma_i - \frac{\sum_{i=1}^{5}\alpha_i}{h_{sur}} \tag{4}$$

Equation 3 is a one -dimensional transient heat conduction equation with a time dependent source term. It has been solved by the method of expansion of eigenfunctions, the procedure for which is described ahead. First, we solve equation 3 by method of separation of variables neglecting the source term. Assuming a product solution of the form $V = F(x) \cdot G(t)$ and feeding it in equation 3, neglecting the source term for the time being

$$F''G = \frac{\rho c_p}{k} FG'$$

$$\Rightarrow \frac{F''}{F} = \frac{\rho c_p}{k} \frac{G'}{G} \tag{5}$$

Since left hand side is a function of x alone and right-hand side of t alone, they can be equal to each other only if each is equal to a constant. That constant can be either positive, zero or negative. It can be seen that if this constant is 0, then the function G will be a constant. This will mean that V and thus, T is independent of time which is not physically meaningful. If that constant is positive, then G shall be an increasing function of time. This is again not physically meaningful. Therefore, the only possibility consistent with the physical nature of the problem is for that constant to be negative. Let this constant be $-p^2$

$$\frac{F''}{F} = \frac{\rho c_p}{k} \frac{G'}{G} = -p^2$$

$$\Rightarrow F'' + p^2 F = 0 \text{ and } \frac{G'}{G} = -\frac{kp^2}{\rho c_p} \tag{6}$$

The next step is to solve the spatial ordinary differential equation out of the two obtained above. It is a standard 2nd order ordinary differential equation with constant coefficients, whose general solution is

$$F = C_1 \cos(px) + C_2 \sin(px)$$

(7)

The temporal part of the general solution needs to be assumed as an unknown function at this point, say *R(t)*. Thus, the general solution for V becomes

$$V = \{C_1 \cos(px) + C_2 \sin(px)\} R(t)$$

(8)

Applying the boundary conditions yields

$$C_2 = 0$$

(9)

$$p \tan(pH) = \frac{h_{sur}}{k}$$

(10)

Equation 10 suggests that there are infinitely many possible values of p. Hence, the general solution of V will be a sum of all possible solutions, one for each value of p.

$$V = \sum_{n=1}^{\infty} C_{1-n} \cos(p_n x).R_n(t)$$

(11)

Expressing the source term in Equation 3 as a generalized fourier cosine series

$$\frac{\rho c_p}{k}\left[\sum_{i=1}^{5}\gamma_i\mu_i e^{\mu_i t}+\frac{\sum_{i=1}^{5}\alpha_i\beta_i e^{\beta_i t}}{h_{sur}}\right]=\sum_{n=1}^{\infty}S_n\cos\left(p_n x\right)$$

where

$$S_n=\frac{\displaystyle\int_0^H\frac{\rho c_p}{k}\left[\sum_{i=1}^{5}\gamma_i\mu_i e^{\mu_i t}+\frac{\sum_{i=1}^{5}\alpha_i\beta_i e^{\beta_i t}}{h_{sur}}\right].\cos\left(p_n x\right).dx}{\displaystyle\int_0^H\cos^2\left(p_n x\right).dx}$$

$$\Rightarrow S_n=\frac{\rho c_p}{k}\left[\sum_{i=1}^{5}\gamma_i\mu_i e^{\mu_i t}+\frac{\sum_{i=1}^{5}\alpha_i\beta_i e^{\beta_i t}}{h_{sur}}\right]\frac{\dfrac{\sin\left(p_n H\right)}{p_n}}{\dfrac{H}{2}+\dfrac{\sin\left(2p_n H\right)}{4p_n}} \tag{12}$$

Substituting equations 11 and 12 in 3

$$\sum_{n=1}^{\infty}-p_n^2 C_{1-n}\cos\left(p_n x\right)R_n(t)-\sum_{n=1}^{\infty}S_n\cos\left(p_n x\right)$$

$$=\frac{\rho c_p}{k}\sum_{n=1}^{\infty}C_{1-n}\cos\left(p_n x\right)\frac{dR_n(t)}{dt}$$

$$\Rightarrow\sum_{n=1}^{\infty}\left[-p_n^2 C_{1-n}R_n(t)-S_n-\frac{\rho c_p}{k}C_{1-n}\frac{dR_n(t)}{dt}\right]\cos\left(p_n x\right)=0 \tag{13}$$

Since LHS=0 for all *x*, therefore

$$-p_n^2 C_{1-n}R_n(t)-S_n-\frac{\rho c_p}{k}C_{1-n}\frac{dR_n(t)}{dt}=0$$

$$\Rightarrow\frac{dR_n(t)}{dt}+\frac{p_n^2 k R_n(t)}{\rho c_p}=-\frac{S_n k}{\rho c_p C_{1-n}} \tag{14}$$

Equation 14 is a first order linear ordinary differential equation in $R_n(t)$. It can be solved by the method of integrating factor as

$$R_n e^{\frac{p_n^2 kt}{\rho c_p}} = \int -\frac{S_n k e^{\frac{p_n^2 kt}{\rho c_p}}}{\rho c_p C_{1-n}} \cdot dt + J_n$$

$$\Rightarrow R_n e^{\frac{p_n^2 kt}{\rho c_p}} = \int -\frac{\rho c_p}{k}\left[\sum_{i=1}^{5}\gamma_i \mu_i e^{\mu_i t} + \frac{\sum_{i=1}^{5}\alpha_i \beta_i e^{\beta_i t}}{h_{sur}}\right]\frac{\dfrac{\sin(p_n H)}{p_n}}{\dfrac{H}{2}+\dfrac{\sin(2p_n H)}{4p_n}}\frac{k e^{\frac{p_n^2 kt}{\rho c_p}}}{\rho c_p C_{1-n}}\cdot dt + J_n$$

$$\Rightarrow R_n e^{\frac{p_n^2 kt}{\rho c_p}} = -\frac{\dfrac{\sin(p_n H)}{p_n}}{\dfrac{H}{2}+\dfrac{\sin(2p_n H)}{4p_n}}\cdot\frac{1}{C_{1-n}}\int\left[\sum_{i=1}^{5}\gamma_i \mu_i e^{\mu_i t} + \frac{\sum_{i=1}^{5}\alpha_i \beta_i e^{\beta_i t}}{h_{sur}}\right]e^{\frac{p_n^2 kt}{\rho c_p}}\cdot dt + J_n \qquad (15)$$

Simplifying

$$R_n e^{\frac{p_n^2 kt}{\rho c_p}} = -\frac{\dfrac{\sin(p_n H)}{p_n}}{\dfrac{H}{2}+\dfrac{\sin(2p_n H)}{4p_n}}\cdot\frac{1}{C_{1-n}}\int\left[e^{\frac{p_n^2 kt}{\rho c_p}}\sum_{i=1}^{5}\gamma_i \mu_i e^{\mu_i t} + e^{\frac{p_n^2 kt}{\rho c_p}}\frac{\sum_{i=1}^{5}\alpha_i \beta_i e^{\beta_i t}}{h_{sur}}\right]\cdot dt + J_n$$

$$\Rightarrow R_n e^{\frac{p_n^2 kt}{\rho c_p}} = -\frac{\dfrac{\sin(p_n H)}{p_n}}{\dfrac{H}{2}+\dfrac{\sin(2p_n H)}{4p_n}}\cdot\frac{1}{C_{1-n}}\int\left[\sum_{i=1}^{5}\gamma_i \mu_i e^{\frac{p_n^2 kt}{\rho c_p}} e^{\mu_i t} + \frac{\sum_{i=1}^{5}\alpha_i \beta_i e^{\frac{p_n^2 kt}{\rho c_p}} e^{\beta_i t}}{h_{sur}}\right]\cdot dt + J_n$$

$$\Rightarrow R_n e^{\frac{p_n^2 kt}{\rho c_p}} = -\frac{\dfrac{\sin(p_n H)}{p_n}}{\dfrac{H}{2}+\dfrac{\sin(2p_n H)}{4p_n}}\cdot\frac{1}{C_{1-n}}\int\left[\sum_{i=1}^{5}\gamma_i \mu_i e^{\left(\frac{p_n^2 k}{\rho c_p}+\mu_i\right)t} + \frac{\sum_{i=1}^{5}\alpha_i \beta_i e^{\left(\frac{p_n^2 k}{\rho c_p}+\beta_i\right)t}}{h_{sur}}\right]\cdot dt + J_n$$

$$\Rightarrow R_n e^{\frac{p_n^2 kt}{\rho c_p}} = -\frac{\dfrac{\sin(p_n H)}{p_n}}{\dfrac{H}{2}+\dfrac{\sin(2p_n H)}{4p_n}} \cdot \frac{1}{C_{1-n}}\left[\sum_{i=1}^{5}\int \gamma_i \mu_i e^{\left(\frac{p_n^2 k}{\rho c_p}+\mu_i\right)t}\cdot dt + \frac{\displaystyle\sum_{i=1}^{5}\int \alpha_i \beta_i e^{\left(\frac{p_n^2 k}{\rho c_p}+\beta_i\right)t}\cdot dt}{h_{sur}}\right]+J_n$$

$$\Rightarrow R_n e^{\frac{p_n^2 kt}{\rho c_p}} = -\frac{\dfrac{\sin(p_n H)}{p_n}}{\dfrac{H}{2}+\dfrac{\sin(2p_n H)}{4p_n}} \cdot \frac{1}{C_{1-n}}\left[\sum_{i=1}^{5}\left(\frac{\gamma_i \mu_i e^{\left(\frac{p_n^2 k}{\rho c_p}+\mu_i\right)t}}{\dfrac{p_n^2 k}{\rho c_p}+\mu_i}\right)+\frac{\displaystyle\sum_{i=1}^{5}\left(\frac{\alpha_i \beta_i e^{\left(\frac{p_n^2 k}{\rho c_p}+\beta_i\right)t}}{\dfrac{p_n^2 k}{\rho c_p}+\beta_i}\right)}{h_{sur}}\right]+J_n$$

$$\Rightarrow R_n e^{\frac{p_n^2 kt}{\rho c_p}} = -\frac{\dfrac{\sin(p_n H)}{p_n}}{\dfrac{H}{2}+\dfrac{\sin(2p_n H)}{4p_n}} \cdot \frac{e^{\frac{p_n^2 kt}{\rho c_p}}}{C_{1-n}}\left[\sum_{i=1}^{5}\left(\frac{\gamma_i \mu_i e^{\mu_i t}}{\dfrac{p_n^2 k}{\rho c_p}+\mu_i}\right)+\frac{\displaystyle\sum_{i=1}^{5}\left(\frac{\alpha_i \beta_i e^{\beta_i t}}{\dfrac{p_n^2 k}{\rho c_p}+\beta_i}\right)}{h_{sur}}\right]+J_n$$

$$\Rightarrow R_n = \left[J_n e^{-\frac{p_n^2 kt}{\rho c_p}}\right]-\left[\left\{\frac{4\sin(p_n H)}{C_{1-n}\{2p_n H+\sin(2p_n H)\}}\right\}\left\{\sum_{i=1}^{5}\left(\frac{\gamma_i \mu_i e^{\mu_i t}}{\dfrac{p_n^2 k}{\rho c_p}+\mu_i}\right)+\frac{\displaystyle\sum_{i=1}^{5}\left(\frac{\alpha_i \beta_i e^{\beta_i t}}{\dfrac{p_n^2 k}{\rho c_p}+\beta_i}\right)}{h_{sur}}\right\}\right] \tag{16}$$

Feeding Equation 16 in 11

$$V = \sum_{n=1}^{\infty}\left[Y_n \cos(p_n x) e^{-\frac{p_n^2 k t}{\rho c_p}} \right] - \left[\left\{ \frac{4\cos(p_n x)\sin(p_n H)}{\{2p_n H + \sin(2p_n H)\}} \right\} \left\{ \sum_{i=1}^{5}\left(\frac{\gamma_i \mu_i e^{\mu_i t}}{\frac{p_n^2 k}{\rho c_p} + \mu_i} \right) + \frac{\sum_{i=1}^{5}\left(\frac{\alpha_i \beta_i e^{\beta_i t}}{\frac{p_n^2 k}{\rho c_p} + \beta_i} \right)}{h_{sur}} \right\} \right] \tag{17}$$

Where $Y_n = C_{1-n} J_n$. The last step is to use the initial condition and evaluate the unknown Y_n. Using Equations 4 and 17

$$\sum_{n=1}^{\infty}\left[Y_n \cos(p_n x) \right] - \left[\left\{ \frac{4\cos(p_n x)\sin(p_n H)}{\{2p_n H + \sin(2p_n H)\}} \right\} \left\{ \sum_{i=1}^{5}\left(\frac{\gamma_i \mu_i}{\frac{p_n^2 k}{\rho c_p} + \mu_i} \right) + \frac{\sum_{i=1}^{5}\left(\frac{\alpha_i \beta_i}{\frac{p_n^2 k}{\rho c_p} + \beta_i} \right)}{h_{sur}} \right\} \right]$$

$$= T_0 - \sum_{i=1}^{5}\gamma_i - \frac{\sum_{i=1}^{5}\alpha_i}{h_{sur}} \tag{18}$$

Expressing right hand side of Equation 18 as a generalized fourier cosine series

$$\sum_{n=1}^{\infty}\left[Y_n \cos(p_n x) \right] - \left[\left\{ \frac{4\cos(p_n x)\sin(p_n H)}{\{2p_n H + \sin(2p_n H)\}} \right\} \left\{ \sum_{i=1}^{5}\left(\frac{\gamma_i \mu_i}{\frac{p_n^2 k}{\rho c_p} + \mu_i} \right) + \frac{\sum_{i=1}^{5}\left(\frac{\alpha_i \beta_i}{\frac{p_n^2 k}{\rho c_p} + \beta_i} \right)}{h_{sur}} \right\} \right]$$

$$= \sum_{n=1}^{\infty} Z_n \cos(p_n x) \tag{19}$$

where

$$Z_n = \frac{\int_0^H \left[T_0 - \sum_{i=1}^{5} \gamma_i - \frac{\sum_{i=1}^{5} \alpha_i}{h_{sur}} \right] \cos(p_n x)\cdot dx}{\int_0^H \cos^2(p_n x)\cdot dx}$$

$$\Rightarrow Z_n = \left[T_0 - \sum_{i=1}^{5} \gamma_i - \frac{\sum_{i=1}^{5} \alpha_i}{h_{sur}} \right]\cdot \left[\frac{\dfrac{\sin(p_n H)}{p_n}}{\dfrac{H}{2} + \dfrac{\sin(2 p_n H)}{4 p_n}} \right] \tag{20}$$

$$\sum_{n=1}^{\infty} Y_n \cos(p_n x) - \sum_{n=1}^{\infty} \left[\left\{ \frac{4\cos(p_n x)\sin(p_n H)}{\{2 p_n H + \sin(2 p_n H)\}} \right\} \left\{ \sum_{i=1}^{5} \left(\frac{\gamma_i \mu_i}{\dfrac{p_n^2 k}{\rho c_p} + \mu_i} \right) + \frac{\sum_{i=1}^{5}\left(\dfrac{\alpha_i \beta_i}{\dfrac{p_n^2 k}{\rho c_p} + \beta_i} \right)}{h_{sur}} \right\} \right] - \sum_{n=1}^{\infty} Z_n \cos(p_n x) = 0$$

$$\sum_{n=1}^{\infty} \left[[Y_n] - [Z_n] - \left[\left\{ \frac{4\sin(p_n H)}{\{2 p_n H + \sin(2 p_n H)\}} \right\} \left\{ \sum_{i=1}^{5} \left(\frac{\gamma_i \mu_i}{\dfrac{p_n^2 k}{\rho c_p} + \mu_i} \right) + \frac{1}{h_{sur}} \sum_{i=1}^{5} \left(\frac{\alpha_i \beta_i}{\dfrac{p_n^2 k}{\rho c_p} + \beta_i} \right) \right\} \right] \right] \cos(p_n x) = 0 \tag{21}$$

Since LHS=0 for all x, therefore

$$[Y_n] - [Z_n] - \left[\left\{ \frac{4\sin(p\,H)}{\{2 p_n H\ \sin(2 p_n H)\}} \right\} \left\{ \sum_{i=1}^{5} \left(\frac{\gamma_i \mu_i}{\dfrac{p_n^2 k}{\rho c_p} + \mu_i} \right) + \frac{1}{sur} \sum_{i=1}^{5} \left(\frac{\alpha_i \beta_i}{\dfrac{p_n^2 k}{\rho c_p} + \beta_i} \right) \right\} \right] =$$

$$\Rightarrow Y_n = Z_n + \left[\left\{ \frac{4\sin(p\,H)}{\{2 p_n H\ \sin(2 p_n H)\}} \right\} \left\{ \sum_{i=1}^{5} \left(\frac{\gamma_i \mu_i}{p\ k} \right)\ \frac{1}{sur} \sum_{i=1}^{5} \left(\frac{\alpha_i \beta_i}{p\ k} \right) \right\} \right] \tag{22}$$

Feeding equation 22 in 17 completes the general solution for *V(x.t)*

$$V = \sum_{n=1}^{\infty} Z_n \cos(p_n x) e^{-\frac{p_n^2 kt}{\rho c_p}} + \sum_{n=1}^{\infty} \left[\left\{ \frac{4\sin(p_n H)\cos(p_n x) e^{-\frac{p_n^2 kt}{\rho c_p}}}{\{2p_n H + \sin(2p_n H)\}} \right\} \left(\sum_{i=1}^{5} \left(\frac{\gamma_i \mu_i}{\frac{p_n^2 k}{\rho c_p} + \mu_i} \right) + \frac{1}{h_{sur}} \sum_{i=1}^{5} \left(\frac{\alpha_i \beta_i}{\frac{p_n^2 k}{\rho c_p} + \beta_i} \right) \right) \right]$$

$$- \sum_{n=1}^{\infty} \left[\left\{ \frac{4\cos(p_n x)\sin(p_n H)}{\{2p_n H + \sin(2p_n H)\}} \right\} \left\{ \sum_{i=1}^{5} \left(\frac{\gamma_i \mu_i e^{\mu_i t}}{\frac{p_n^2 k}{\rho c_p} + \mu_i} \right) + \frac{\sum_{i=1}^{5} \left(\frac{\alpha_i \beta_i e^{\beta_i t}}{\frac{p_n^2 k}{\rho c_p} + \beta_i} \right)}{h_{sur}} \right\} \right]$$

$$(23)$$

Thus, the final solution for required temperature field becomes

$$T = \sum_{i=1}^{5} \gamma_i e^{\mu_i t} + \frac{\sum_{i=1}^{5} \alpha_i e^{\beta_i t}}{h_{sur}} + \sum_{n=1}^{\infty} Z_n \cos(p_n x) e^{-\frac{p_n^2 kt}{\rho c_p}} +$$

$$\sum_{n=1}^{\infty} \left[\left\{ \frac{4\sin(p_n H)\cos(p_n x) e^{-\frac{p_n^2 kt}{\rho c_p}}}{\{2p_n H + \sin(2p_n H)\}} \right\} \left\{ \sum_{i=1}^{5} \left(\frac{\gamma_i \mu_i}{\frac{p_n^2 k}{\rho c_p} + \mu_i} \right) + \frac{1}{h_{sur}} \sum_{i=1}^{5} \left(\frac{\alpha_i \beta_i}{\frac{p_n^2 k}{\rho c_p} + \beta_i} \right) \right\} \right] -$$

$$\sum_{n=1}^{\infty} \left[\left\{ \frac{4\cos(p_n x)\sin(p_n H)}{\{2p_n H + \sin(2p_n H)\}} \right\} \left\{ \sum_{i=1}^{5} \left(\frac{\gamma_i \mu_i e^{\mu_i t}}{\frac{p_n^2 k}{\rho c_p} + \mu_i} \right) + \frac{\sum_{i=1}^{5} \left(\frac{\alpha_i \beta_i e^{\beta_i t}}{\frac{p_n^2 k}{\rho c_p} + \beta_i} \right)}{h_{sur}} \right\} \right]$$

$$(24)$$

Expanding back Z_n

$$T = \sum_{i=1}^{5} \gamma_i e^{\mu_i t} + \frac{1}{h_{sur}} \sum_{i=1}^{5} \alpha_i e^{\beta_i t} + \sum_{n=1}^{\infty} \left[T_0 - \sum_{i=1}^{5} \gamma_i - \frac{\sum_{i=1}^{5} \alpha_i}{h_{sur}} \right] \cdot \left[\frac{\dfrac{\sin(p_n H)}{p_n}}{\dfrac{H}{2} + \dfrac{\sin(2 p_n H)}{4 p_n}} \right] \cos(p_n x) e^{-\frac{p_n^2 kt}{\rho c_p}} +$$

$$\sum_{n=1}^{\infty} \left[\left\{ \frac{4 \sin(p_n H) \cos(p_n x) e^{-\frac{p_n^2 kt}{\rho c_p}}}{\{2 p_n H + \sin(2 p_n H)\}} \right\} \left\{ \sum_{i=1}^{5} \left(\frac{\gamma_i \mu_i}{\dfrac{p_n^2 k}{\rho c_p} + \mu_i} \right) + \frac{1}{h_{sur}} \cdot \sum_{i=1}^{5} \left(\frac{\alpha_i \beta_i}{\dfrac{p_n^2 k}{\rho c_p} + \beta_i} \right) \right\} \right] -$$

$$\sum_{n=1}^{\infty} \left[\left\{ \frac{4 \cos(p_n x) \sin(p_n H)}{\{2 p_n H + \sin(2 p_n H)\}} \right\} \left\{ \sum_{i=1}^{5} \left(\frac{\gamma_i \mu_i e^{\mu_i t}}{\dfrac{p_n^2 k}{\rho c_p} + \mu_i} \right) + \frac{1}{h_{sur}} \cdot \sum_{i=1}^{5} \left(\frac{\alpha_i \beta_i e^{\beta_i t}}{\dfrac{p_n^2 k}{\rho c_p} + \beta_i} \right) \right\} \right]$$

where roots of the equation $p \tan(pH) = \dfrac{h_{sur}}{k}$ gives the values of p_n $(n = 1, 2, 3 \ldots \ldots)$

$$(25)$$

Question 20

In many countries, the electricity cost during the night time is lesser than that during day hours. This disparity led to the innovation called electric thermal storage heater. It consists of a thin block made of a material of large specific heat capacity. An electrical resistor is placed inside it. During night time, when electricity expenses are low, current is passed through the resistor. Heat generated by it warms the block. During day, when electricity cost is high, room air is heated by passing it over the block and is circulated indoors. Thus, space heating is accomplished during day at the cost of night time electric energy and so this device helps save money. Consider the shown electric thermal storage heater. Suppose it gets charged for a period t_0 and the power of the heater is P. Find temperature distribution in the block at the end of charging period. Thermal conductivity, specific heat capacity and mass density of the block are k, c_p, ρ. The block is perfectly insulated from the surroundings. Initial temperature of the block is T_a (room air temperature).

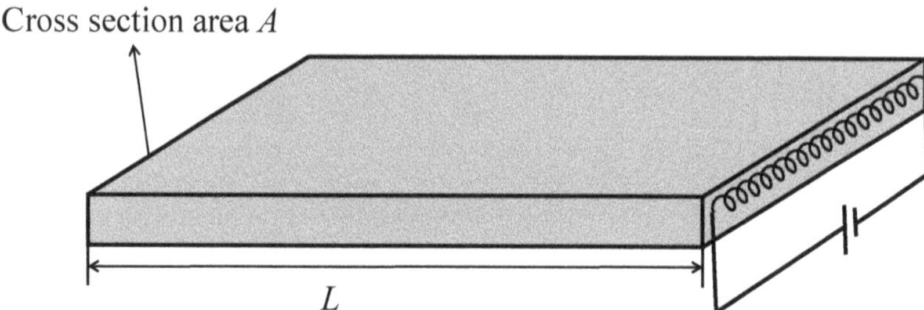

Solution

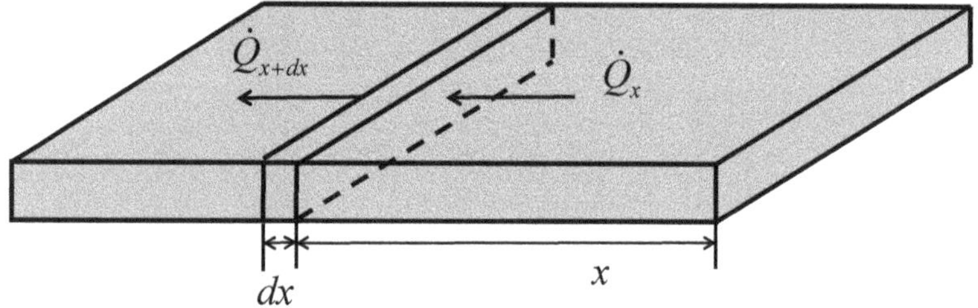

Applying law of energy conservation for the differential element shown

$$\dot{Q}_x - \dot{Q}_{x+dx} = \frac{\partial}{\partial t}\left(E_{int\,ernal}\right)$$

$$\Rightarrow \dot{Q}_x - \left(\dot{Q}_x + \frac{\partial \dot{Q}_x}{\partial x}dx\right) = \frac{\partial}{\partial t}\left(E_{int\,ernal}\right)$$

$$\Rightarrow -\frac{\partial}{\partial x}\left(-kA\frac{\partial T}{\partial x}\right)dx = \frac{\partial}{\partial t}\left(\rho A \cdot dx \cdot c_p T\right)$$

$$\Rightarrow \frac{\partial^2 T}{\partial x^2} = \frac{\rho c_p}{k}\frac{\partial T}{\partial t} \tag{1}$$

The relevant boundary and initial conditions are:

$$\text{B.C 1}: \quad -kA\frac{\partial T}{\partial x}(0,t) = P$$

$$\text{B.C 2}: \quad \frac{\partial T}{\partial x}(L,t) = 0$$

$$\text{I.C}: \quad T(x,0) = T_a \tag{2}$$

Let $T = v + C_1 x^2 + C_2 x$. Under this change of variable, equations 1 and 2 transform to

$$\frac{\partial^2 v}{\partial x^2} + 2C_1 = \frac{\rho c_p}{k}\frac{\partial v}{\partial t} \tag{3}$$

$$\text{B.C 1}: \quad \frac{\partial v}{\partial x}(0,t) + C_2 = -\frac{P}{kA}$$

$$\text{B.C 2}: \quad \frac{\partial v}{\partial x}(L,t) + 2C_1 L + C_2 = 0$$

$$\text{I.C}: \quad v(x,0) = T_a - C_1 x^2 - C_2 x \tag{4}$$

Choosing $C_2 = -\dfrac{P}{kA}$ and $C_1 = \dfrac{P}{2kAL}$, and defining $\dfrac{\rho c_p}{k}$ as a, the problem becomes

$$\frac{\partial^2 v}{\partial x^2} + \frac{P}{kAL} = a\frac{\partial v}{\partial t} \tag{5}$$

$$\text{B.C 1}: \quad \frac{\partial v}{\partial x}(0,t) = 0$$

$$\text{B.C 2}: \quad \frac{\partial v}{\partial x}(L,t) = 0$$

$$\text{I.C}: \quad v(x,0) = T_a - \frac{P}{2kAL}x^2 + \frac{P}{kA}x \tag{6}$$

Equation 5 is a transient one-dimensional heat conduction equation with a constant source term. It can be solved by the method of expansion of eigen functions, the procedure for which is described ahead. The first step is to ignore the source term and decompose the remaining partial differential equation into two ordinary differential equations by the method of separation of variables. Assuming a product solution of the form $v = F(x) \cdot G(t)$ and feeding it in equation 5, neglecting the source term for the time being

$$F''G = aFG'$$

$$\Rightarrow \frac{F''}{F} = a\frac{G'}{G} \tag{7}$$

Since left hand side is a function of x alone and right-hand side of t alone, they can be equal to each other only if each is equal to a constant. That constant can be either positive, zero or negative. It can be seen that if this constant is 0, then the function G will be a constant. This will mean that v and thus, T is independent of time which is not physically meaningful. If that constant is positive, then G shall be an increasing function of time. This is again not physically meaningful because after a certain time, steady state will be attained and so the temporal part of the solution must decrease with time. Therefore, the only possibility consistent with the physical nature of the problem is for that constant to be negative. Let this constant be $-w^2$

$$\frac{F''}{F} = a\frac{G'}{G} = -w^2$$

$$\Rightarrow F'' + w^2 F = 0 \text{ and } \frac{G'}{G} = -\frac{w^2}{a} \tag{8}$$

The next step is to solve the spatial ordinary differential equation out of the two obtained above. It is a standard 2nd order ordinary differential equation with constant coefficients, whose general solution is

$$F = C_{11}\cos(wx) + C_{22}\sin(wx) \tag{9}$$

The temporal part of the general solution needs to be assumed as an unknown function at this point, say *R(t)*. Thus, the general solution for v becomes

$$v = \{C_{11}\cos(wx) + C_{22}\sin(wx)\}R(t) \tag{10}$$

Using the two boundary conditions given in equation 6

$$C_{22} = 0 \tag{11}$$

$$\sin(wL) = 0$$
$$\Rightarrow w = \frac{n\pi}{L}, n = 1,2,3\ldots\ldots \tag{12}$$

Equation 12 suggests that there are infinitely many possible values of w. Hence, the general solution of v will be a sum of all possible solutions, one for each value of w.

$$v = \sum_{n=1}^{\infty}\cos(w_n x)C_{11-n}R_n(t) \tag{13}$$

Expressing the source term in Eq. (5) as a generalized fourier cosine series

$$\frac{P}{kAL} = \sum_{n=1}^{\infty} H_n \cos(w_n x)$$

where

$$H_n = \frac{\displaystyle\int_0^L \frac{P}{kAL}.\cos(w_n x).dx}{\displaystyle\int_0^L \cos^2(w_n x).dx}$$

$$\Rightarrow H_n = \frac{P}{kAL}\left\{\frac{\dfrac{\sin(w_n L)}{w_n}}{\dfrac{L}{2} + \dfrac{\sin(2w_n L)}{4w_n}}\right\} \tag{14}$$

Substituting equations 13 and 14 in 5

$$-\sum_{n=1}^{\infty} w_n^2 \cos(w_n x) C_{11-n} R_n(t) + \sum_{n=1}^{\infty} H_n \cos(w_n x) = a\sum_{n=1}^{\infty} \cos(w_n x) C_{11-n} \frac{dR_n(t)}{dt}$$

$$\Rightarrow \sum_{n=1}^{\infty}\left[-w_n^2 C_{11-n} R_n(t) + H_n - aC_{11-n}\frac{dR_n(t)}{dt}\right]\cos(w_n x) = 0 \tag{15}$$

Since LHS = 0 for all x, therefore

$$-w_n^2 C_{11-n} R_n(t) + H_n - aC_{11-n}\frac{dR_n(t)}{dt} = 0$$

$$\rightarrow \frac{dR_n(t)}{dt} + \frac{w_n^2 R_n(t)}{a} = \frac{H_n}{aC_{11-n}} \tag{16}$$

Equation 16 is a first order linear ordinary differential equation in *Rn(t)*. It can be solved by the method of integrating factor as

$$R_n e^{\frac{w_n^2 t}{a}} = \int \frac{H_n}{aC_{11-n}} e^{\frac{w_n^2 t}{a}} \cdot dt + J_n$$

$$\Rightarrow R_n = \frac{H_n}{w_n^2 C_{11-n}} + J_n e^{-\frac{w_n^2 t}{a}} \tag{17}$$

Feeding Equation 17 in 13

$$v = \sum_{n=1}^{\infty} \cos(w_n x) \left[\frac{H_n}{w_n^2} + Q_n e^{-\frac{w_n^2 t}{a}} \right] \tag{18}$$

Where $Q_n = C_{11-n} J_n$. The last step is to use the initial condition and evaluate the unknown Q_n. Using Equations 6 and 18

$$\sum_{n=1}^{\infty} \cos(w_n x) \left[\frac{H_n}{w_n^2} + Q_n \right] = T_a - \frac{P}{2kAL} x^2 + \frac{P}{kA} x \tag{19}$$

Expressing right hand side of Equation 19 as a generalized fourier cosine series

$$\sum_{n=1}^{\infty} \cos(w_n x) \left[\frac{H_n}{w_n^2} + Q_n \right] = \sum_{n=1}^{\infty} S_n \cos(w_n x)$$

$$\Rightarrow \sum_{n=1}^{\infty} \cos(w_n x) \left[\frac{H_n}{w_n^2} + Q_n - S_n \right] = 0$$

$$\Rightarrow Q_n = S_n - \frac{H_n}{w_n^2} \tag{20}$$

Where

$$S_n = \frac{\displaystyle\int_0^L \left\{ T_a - \frac{P}{2kAL}x^2 + \frac{P}{kA}x \right\} \cos(w_n x).dx}{\displaystyle\int_0^L \cos^2(w_n x).dx}$$

$$= \frac{\left\{ \left(T_a + \dfrac{PL}{2kA} \right)\left(\dfrac{\sin(w_n L)}{w_n} \right) \right\} - \left\{ \dfrac{P}{kAw_n^2} \right\} + \left\{ \dfrac{P\sin(w_n L)}{kALw_n^3} \right\}}{\dfrac{L}{2} + \dfrac{\sin(2w_n L)}{4w_n}} \tag{21}$$

The numerator integral in equation 21 has been evaluated by integration by parts while the denominator integral is straightforward. The general solution for *v(x,t)* therefore becomes

$$v = \sum_{n=1}^{\infty} \cos(w_n x)\left[\frac{H_n}{w_n^2} + \left\{ S_n - \frac{H_n}{w_n^2} \right\} e^{-\frac{w_n^2 t}{a}} \right] \tag{22}$$

Therefore, the general solution for *T(x,t)* is

$$T = \frac{P}{2kAL}x^2 - \frac{P}{kA}x + \sum_{n=1}^{\infty} \cos(w_n x)\left[\frac{H_n}{w_n^2} + \left\{ S_n - \frac{H_n}{w_n^2} \right\} e^{-\frac{w_n^2 t}{a}} \right] \tag{23}$$

Expanding back H_n, S_n

$$T = \frac{P}{2kAL}x^2 - \frac{P}{kA}x + \sum_{n=1}^{\infty}\cos(w_n x)\left[\begin{array}{c}\left[\dfrac{\dfrac{P\sin(w_n L)}{kALw_n^3}}{\dfrac{L}{2}+\dfrac{\sin(2w_n L)}{4w_n}}\right]+\\[2em] \dfrac{\left\{\left(T_a + \dfrac{PL}{2kA}\right)\left(\dfrac{\sin(w_n L)}{w_n}\right)\right\}e^{-\frac{w_n^2 t}{a}} - \left\{\dfrac{P}{kAw_n^2}\right\}e^{-\frac{w_n^2 t}{a}} + \left\{\dfrac{P\sin(w_n L)}{kALw_n^3}\right\}e^{-\frac{w_n^2 t}{a}}}{\dfrac{L}{2}+\dfrac{\sin(2w_n L)}{4w_n}} - \\[2em] \dfrac{P\sin(w_n L)e^{-\frac{w_n^2 t}{a}}}{kALw_n^3}\bigg/\left(\dfrac{L}{2}+\dfrac{\sin(2w_n L)}{4w_n}\right)\end{array}\right] \tag{24}$$

Expanding back w_n, a

$$T = \frac{P}{2kAL}x^2 - \frac{P}{kA}x - \frac{2PL}{kA\pi^2}\sum_{n=1}^{\infty}\cos\left(\frac{n\pi x}{L}\right)\frac{e^{-\frac{kn^2\pi^2 t}{L^2\rho c_p}}}{n^2} \tag{25}$$

The temperature distribution at end of charging period will be $T(x,t_0)$

$$T = \frac{P}{2kAL}x^2 - \frac{P}{kA}x - \frac{2PL}{kA\pi^2}\sum_{n=1}^{\infty}\cos\left(\frac{n\pi x}{L}\right)\frac{e^{-\frac{kn^2\pi^2 t_0}{L^2\rho c_p}}}{n^2} \tag{26}$$

Question 21

A solar air heater is a device used for heating air via solar energy to be used for drying and space heating purposes. It consists of a channel for air flow, well insulated at the base and sides but covered by a glass cover at the top. Air enters from one end, and gets heated by solar radiation as it flows. It is not possible to use this setup at night. To combat this problem, the concept of thermal storage is used. The solar air heater is provided with a heat storage medium, which stores some of the incident solar energy during day and releases it at night. This modification is known as a packed bed solar air heater. Shown is a packed bed solar air heater used in alternate charging and discharging mode. It consists of a bed of rocks placed in a cuboidal channel well insulated at the base and side walls, open at the top. The setup is placed in open sunlight during daytime. At the end of the day, the entire rock bed heats upto a temperature T_0. [The height of rock bed is small such that vertical temperature gradients inside it can be ignored]. After sunset, the top is covered by a perfect insulation cover. Air is pumped in from one end. It enters at instantaneous ambient temperature. Take $t=0$ as the moment air is begun flowing in. Ambient temperature for the night time period can be described by the function: $T_a = \sum_{i=1}^{5} \gamma_i e^{\mu_i t}$.

Considering the heat transfer coefficient between air and rocks to be infinite, the local temperatures of air and rocks coincide. The governing equation of heat transfer for the system can be derived using energy conservation for a differential element and is: $\dfrac{\partial^2 T}{\partial x^2} - \dfrac{\dot{m}_a c_{pa}}{k_s BH} \dfrac{\partial T}{\partial x} = \dfrac{\rho_s c_{ps}}{k_s} \dfrac{\partial T}{\partial t}$

Where T is the common, local temperature of air and rocks, $k_s, \rho_s c_{ps}$ are thermal conductivity, mass density and specific heat capacity of rocks, $\dot{m}_a, c_{pa}$ are mass flow rate and specific heat capacity of air. The relevant boundary and initial conditions

$$\text{B.C 1}: T(0,t) = T_a(t) = \sum_{i=1}^{5} \gamma_i e^{\mu_i t}$$

are: $\text{B.C 2}: \dfrac{\partial T}{\partial x}(L,t) = 0$

$\text{I.C}: T(x,0) = T_0$

Find exit air temperature during discharging period (off-sunshine hours) as a function of time.

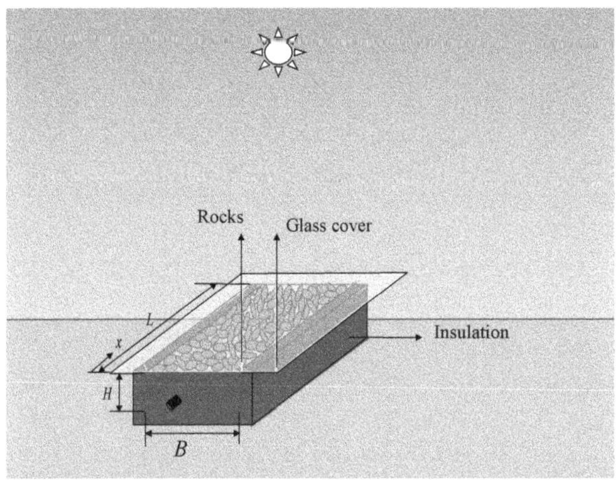

Charging of storage medium (day operation)

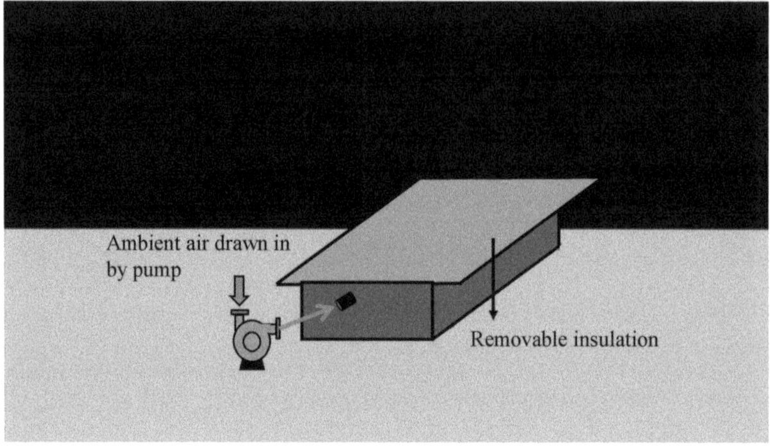

Discharging of storage medium (night operation)

Solution

Let $\theta = T - T_a(t)$. The governing equation and associated boundary, initial conditions then transform to

$$\frac{\partial^2 \theta}{\partial x^2} - a\frac{\partial \theta}{\partial x} = b\frac{\partial \theta}{\partial t} + b\sum_{i=1}^{5}\gamma_i \mu_i e^{\mu_i t} \tag{1}$$

$$\text{B.C 1: } \theta(0,t) = 0$$

$$\text{B.C 2: } \frac{\partial \theta}{\partial x}(L,t) = 0$$

$$\text{I.C : } \theta(x,0) = T_0 - \sum_{i=1}^{5}\gamma_i \tag{2}$$

where $a = \dfrac{\dot{m}_a c_{pa}}{k_s BH}, b = \dfrac{\rho_s c_{ps}}{k_s}$. Equation 1 is an inhomogeneous transient one-dimensional heat conduction equation. It has been solved by the method of expansion of eigen functions. The first step is to ignore the source term in the equation and decompose the remaining PDE into two ODEs by the method of separation of variables. Assuming a product solution of the form $\theta = F(x) \cdot G(t)$ and feeding it in equation 1, ignoring the source term for the time being

$$F''G - aF'G = bFG'$$

$$\Rightarrow \frac{F'' - aF'}{F} = b\frac{G'}{G} \tag{3}$$

Since left hand side is a function of x alone and right-hand side of t alone, they can be equal to each other only if each is equal to a constant. That constant can be either positive, zero or negative. It can be seen that if this constant is 0, then the function G will be a constant. This will mean that θ and thus, T is independent of time which is not physically meaningful. If that constant is positive, then G shall be an increasing function of time. This is again not physically meaningful. Therefore, the only possibility consistent with the physical nature of the problem is for that constant to be negative. Let this constant be $-p^2$

$$\frac{F'' - aF'}{F} = b\frac{G'}{G} = -p^2$$

$$\Rightarrow F'' - aF' + p^2 F = 0 \quad \text{and} \quad \frac{G'}{G} = \frac{-p^2}{b} \tag{4}$$

The next step is to solve the spatial ordinary differential equation out of the two obtained above. It is a standard 2nd order ordinary differential equation with constant coefficients, whose general solution is

$$F = e^{\frac{ax}{2}} \left[C_1 \cos\left(\frac{\sqrt{4p^2 - a^2}}{2} x \right) + C_2 \sin\left(\frac{\sqrt{4p^2 - a^2}}{2} x \right) \right]$$

$$OR \; F = \left\{ C_1 \cos(wx) + C_2 \sin(wx) \right\} e^{\frac{ax}{2}} \tag{5}$$

Where $\dfrac{\sqrt{4p^2 - a^2}}{2} = w$. The temporal part of the general solution needs to be assumed as an unknown function at this point, say *R(t)*. Thus, the general solution for θ becomes

$$\theta = \left\{ C_1 \cos(wx) + C_2 \sin(wx) \right\} e^{\frac{ax}{2}} \cdot R(t) \tag{6}$$

Applying the boundary conditions yields

$$C_1 = 0 \tag{7}$$

$$\tan(wL) = -\frac{2w}{a} \tag{8}$$

Equation 8 suggests that there are infinitely many possible values of *w*. The general solution is thus the sum of all possible solutions (one for each value of *w*)

$$\theta = \sum_{n=1}^{\infty} C_{2-n} \sin(w_n x) e^{\frac{ax}{2}} R_n(t) \tag{9}$$

Feeding equation 9 in 1

$$-\sum_{n=1}^{\infty} C_{2-n} R_n e^{\frac{ax}{2}} \left(\frac{a^2}{4} + w_n^2 \right) \sin\left(w_n x\right) = b \sum_{n=1}^{\infty} C_{2-n} \sin\left(w_n x\right) e^{\frac{ax}{2}} \frac{dR_n}{dt} +$$

$$b \sum_{i=1}^{5} \gamma_i \mu_i\, e^{\mu_i t}$$

$$\Rightarrow -\sum_{n=1}^{\infty} C_{2-n} R_n \left(\frac{a^2}{4} + w_n^2 \right) \sin\left(w_n x\right) = b \sum_{n=1}^{\infty} C_{2-n} \sin\left(w_n x\right) \frac{dR_n}{dt} +$$

$$be^{-\frac{ax}{2}} \sum_{i=1}^{5} \gamma_i \mu_i\, e^{\mu_i t} \tag{10}$$

At this step, the rightmost term on the right-hand side of above equation needs to be expressed as a generalized fourier sine series

$$be^{-\frac{ax}{2}} \sum_{i=1}^{5} \gamma_i \mu_i\, e^{\mu_i t} = \sum_{n=1}^{\infty} J_n \sin\left(w_n x\right)$$

$$\text{where } J_n = \frac{\int_0^L \left\{ be^{-\frac{ax}{2}} \sum_{i=1}^{5} \gamma_i \mu_i\, e^{\mu_i t} \right\} . \sin\left(w_n x\right).dx}{\int_0^L \sin^2\left(w_n x\right).dx}$$

$$\Rightarrow J_n = b\sum_{i=1}^{5} \gamma_i \mu_i\, e^{\mu_i t} \frac{\int_0^L e^{-\frac{ax}{2}} \sin\left(w_n x\right).dx}{\int_0^L \sin^2\left(w_n x\right).dx}$$

$$\Rightarrow J_n = b\sum_{i=1}^{5} \gamma_i \mu_i\, e^{\mu_i t} \left[\frac{\frac{2}{a^2 + 4w_n^2}\left\{ 2w_n - a\sin\left(w_n L\right)e^{-\frac{aL}{2}} - 2w_n \cos\left(w_n L\right)e^{-\frac{aL}{2}} \right\}}{\frac{L}{2} - \frac{\sin\left(2w_n L\right)}{4w_n}} \right] \tag{11}$$

The numerator integral in the expression of J_n has been evaluated using integration by parts while the denominator integral is straightforward. Feeding equation 11 in 10

$$-\sum_{n=1}^{\infty} C_{2-n} R_n \left(\frac{a^2}{4} + w_n^2 \right) \sin\left(w_n x \right) = b \sum_{n=1}^{\infty} C_{2-n} \sin\left(w_n x \right) \frac{dR_n}{dt} + \sum_{n=1}^{\infty} J_n \sin\left(w_n x \right)$$

$$\Rightarrow \sum_{n=1}^{\infty} \left[-C_{2-n} R_n \left(\frac{a^2}{4} + w_n^2 \right) - bC_{2-n} \frac{dR_n}{dt} - J_n \right] \sin\left(w_n x \right) = 0 \tag{12}$$

Since LHS=0 for all *x*, therefore

$$-C_{2-n} R_n \left(\frac{a^2}{4} + w_n^2 \right) - bC_{2-n} \frac{dR_n}{dt} - J_n = 0$$

$$\Rightarrow \frac{dR_n}{dt} + \frac{R_n}{b} \left(\frac{a^2}{4} + w_n^2 \right) = -\frac{J_n}{bC_{2-n}} \tag{13}$$

Equation 13 is a first order linear ordinary differential equation in *Rn(t)*. It can be solved by the method of integrating factor as

$$R_n e^{\frac{t}{b}\left(\frac{a^2}{4} + w_n^2 \right)} = \int -\frac{J_n e^{\frac{t}{b}\left(\frac{a^2}{4} + w_n^2 \right)}}{bC_{2-n}} dt + S_n$$

$$\Rightarrow R_n e^{\frac{t}{b}\left(\frac{a^2}{4} + w_n^2 \right)} = \int -b\sum_{i=1}^{5} \gamma_i \mu_i \, e^{\mu_i t} \left[\frac{\frac{2}{a^2+4w_n^2}\left\{ 2w_n - a\sin\left(w_n L\right)e^{-\frac{aL}{2}} - 2w_n \cos\left(w_n L\right)e^{-\frac{aL}{2}} \right\}}{\frac{L}{2} - \frac{\sin\left(2w_n L\right)}{4w_n}} \right] \frac{e^{\frac{t}{b}\left(\frac{a^2}{4} + w_n^2 \right)}}{bC_{2-n}} dt + S_n$$

$$\Rightarrow R_n e^{\frac{t}{b}\left(\frac{a^2}{4} + w_n^2 \right)} = -\left[\frac{\frac{2}{a^2+4w_n^2}\left\{ 2w_n - a\sin\left(w_n L\right)e^{-\frac{aL}{2}} - 2w_n \cos\left(w_n L\right)e^{-\frac{aL}{2}} \right\}}{C_{2-n}\left\{ \frac{L}{2} - \frac{\sin\left(2w_n L\right)}{4w_n} \right\}} \right] \int \sum_{i=1}^{5} \gamma_i \mu_i \, e^{\mu_i t} e^{\frac{t}{b}\left(\frac{a^2}{4} + w_n^2 \right)} dt + S_n$$

$$\Rightarrow R_n e^{\frac{t}{b}\left(\frac{a^2}{4} + w_n^2 \right)} = -\left[\frac{\frac{2}{a^2+4w_n^2}\left\{ 2w_n - a\sin\left(w_n L\right)e^{-\frac{aL}{2}} - 2w_n \cos\left(w_n L\right)e^{-\frac{aL}{2}} \right\}}{C_{2-n}\left\{ \frac{L}{2} - \frac{\sin\left(2w_n L\right)}{4w_n} \right\}} \right] \sum_{i=1}^{5} \frac{\gamma_i \mu_i e^{\left(\mu_i + \frac{a^2}{4b} + \frac{w_n^2}{b} \right)t}}{\mu_i + \frac{a^2}{4b} + \frac{w_n^2}{b}} + S_n$$

$$\Rightarrow R_n = -\left[\frac{\frac{2}{a^2+4w_n^2}\left\{ 2w_n - a\sin\left(w_n L\right)e^{-\frac{aL}{2}} - 2w_n \cos\left(w_n L\right)e^{-\frac{aL}{2}} \right\}}{C_{2-n}\left\{ \frac{L}{2} - \frac{\sin\left(2w_n L\right)}{4w_n} \right\}} \right] \sum_{i=1}^{5} \frac{\gamma_i \mu_i e^{\mu_i t}}{\mu_i + \frac{a^2}{4b} + \frac{w_n^2}{b}} + S_n e^{-\frac{t}{b}\left(\frac{a^2}{4} + w_n^2 \right)} \tag{14}$$

Feeding equation 14 in 9

$$\theta = \sum_{n=1}^{\infty} \sin(w_n x) e^{\frac{ax}{2}} Z_n e^{-\frac{t}{b}\left(\frac{a^2}{4}+w_n^2\right)}$$

$$-\sum_{n=1}^{\infty} \sin(w_n x) e^{\frac{ax}{2}} \left[\frac{\frac{2}{a^2+4w_n^2}\left\{ 2w_n - a\sin(w_n L)e^{-\frac{aL}{2}} - 2w_n \cos(w_n L)e^{-\frac{aL}{2}} \right\}}{\left\{ \frac{L}{2} - \frac{\sin(2w_n L)}{4w_n} \right\}} \right] \sum_{i=1}^{5} \frac{\gamma_i \mu_i e^{\mu_i t}}{\mu_i + \frac{a^2}{4b} + \frac{w_n^2}{b}} \tag{15}$$

Where $Z_n = C_{2-n} S_n$. The last step is to apply the initial condition and solve for the unknown Z_n. Feeding equation 15 in initial condition mentioned in equation 2

$$T_0 - \sum_{i=1}^{5} \gamma_i = \sum_{n=1}^{\infty} \sin(w_n x) e^{\frac{ax}{2}} Z_n$$

$$-\sum_{n=1}^{\infty} \sin(w_n x) e^{\frac{ax}{2}} \left[\frac{\frac{2}{a^2+4w_n^2}\left\{ 2w_n - a\sin(w_n L)e^{-\frac{aL}{2}} - 2w_n \cos(w_n L)e^{-\frac{aL}{2}} \right\}}{\left\{ \frac{L}{2} - \frac{\sin(2w_n L)}{4w_n} \right\}} \right] \sum_{i=1}^{5} \frac{\gamma_i \mu_i}{\mu_i + \frac{a^2}{4b} + \frac{w_n^2}{b}}$$

$$\Rightarrow \left(T_0 - \sum_{i=1}^{5} \gamma_i \right) e^{-\frac{ax}{2}} = \sum_{n=1}^{\infty} \sin(w_n x) Z_n$$

$$-\sum_{n=1}^{\infty} \sin(w_n x) \left[\frac{\frac{2}{a^2+4w_n^2}\left\{ 2w_n - a\sin(w_n L)e^{-\frac{aL}{2}} - 2w_n \cos(w_n L)e^{-\frac{aL}{2}} \right\}}{\left\{ \frac{L}{2} - \frac{\sin(2w_n L)}{4w_n} \right\}} \right] \sum_{i=1}^{5} \frac{\gamma_i \mu_i}{\mu_i + \frac{a^2}{4b} + \frac{w_n^2}{b}} \tag{16}$$

The left-hand side of above equation can be expressed as a generalized fourier sine series

$$\left(T_0 - \sum_{i=1}^{5}\gamma_i\right)e^{-\frac{ax}{2}} = \sum_{n=1}^{\infty} K_n \sin(w_n x)$$

where

$$K_n = \frac{\displaystyle\int_0^L \left(T_0 - \sum_{i=1}^{5}\gamma_i\right)e^{-\frac{ax}{2}}\sin(w_n x).dx}{\displaystyle\int_0^L \sin^2(w_n x).dx}$$

$$\Rightarrow K_n = \left(T_0 - \sum_{i=1}^{5}\gamma_i\right)\left[\frac{\dfrac{2}{a^2+4w_n^2}\left\{2w_n - a\sin(w_n L)e^{-\frac{aL}{2}} - 2w_n\cos(w_n L)e^{-\frac{aL}{2}}\right\}}{\dfrac{L}{2}-\dfrac{\sin(2w_n L)}{4w_n}}\right] \tag{17}$$

Substituting equation 17 in 16

$$\sum_{n=1}^{\infty} K_n \sin(w_n x) = \sum_{n=1}^{\infty}\sin(w_n x) Z_n$$

$$-\sum_{n=1}^{\infty}\sin(w_n x)\left[\frac{\dfrac{2}{a^2+4w_n^2}\left\{2w_n - a\sin(w_n L)e^{-\frac{aL}{2}} - 2w_n\cos(w_n L)e^{-\frac{aL}{2}}\right\}}{\left\{\dfrac{L}{2}-\dfrac{\sin(2w_n L)}{4w_n}\right\}}\right]\sum_{i=1}^{5}\frac{\gamma_i\mu_i}{\mu_i+\dfrac{a^2}{4b}+\dfrac{w_n^2}{b}}$$

$$\Rightarrow \sum_{n=1}^{\infty}\sin(w_n x)\left[\begin{array}{l}Z_n - \\ \left[\dfrac{\dfrac{2}{a^2+4w_n^2}\left\{2w_n - a\sin(w_n L)e^{-\frac{aL}{2}} - 2w_n\cos(w_n L)e^{-\frac{aL}{2}}\right\}}{\left\{\dfrac{L}{2}-\dfrac{\sin(2w_n L)}{4w_n}\right\}}\right]\sum_{i=1}^{5}\dfrac{\gamma_i\mu_i}{\mu_i+\dfrac{a^2}{4b}+\dfrac{w_n^2}{b}} - \\ K_n \end{array}\right] = 0$$

$$\tag{18}$$

Since LHS =0 for all x, therefore

$$Z_n = K_n + \left[\frac{\dfrac{2}{a^2+4w_n^2}\left\{2w_n - a\sin(w_n L)e^{-\frac{aL}{2}} - 2w_n\cos(w_n L)e^{-\frac{aL}{2}}\right\}}{\left\{\dfrac{L}{2}-\dfrac{\sin(2w_n L)}{4w_n}\right\}}\right]\sum_{i=1}^{5}\frac{\gamma_i\mu_i}{\mu_i+\dfrac{a^2}{4b}+\dfrac{w_n^2}{b}} \tag{19}$$

Substituting equation 19 in 15

$$
\theta = \sum_{n=1}^{\infty} \sin(w_n x)\, e^{\frac{ax}{2}} K_n e^{-\frac{t}{b}\left(\frac{a^2}{4}+w_n^2\right)} + \sum_{n=1}^{\infty} \sin(w_n x)\, e^{\frac{ax}{2}} e^{-\frac{t}{b}\left(\frac{a^2}{4}+w_n^2\right)}
$$

$$
\left[\frac{\dfrac{2}{a^2+4w_n^2}\left\{2w_n - a\sin(w_n L)e^{-\frac{aL}{2}} - 2w_n\cos(w_n L)e^{-\frac{aL}{2}}\right\}}{\left\{\dfrac{L}{2}-\dfrac{\sin(2w_n L)}{4w_n}\right\}}\right] \sum_{i=1}^{5} \frac{\gamma_i \mu_i}{\mu_i + \dfrac{a^2}{4b} + \dfrac{w_n^2}{b}}
$$

$$
-\sum_{n=1}^{\infty} \sin(w_n x)\, e^{\frac{ax}{2}} \left[\frac{\dfrac{2}{a^2+4w_n^2}\left\{2w_n - a\sin(w_n L)e^{-\frac{aL}{2}} - 2w_n\cos(w_n L)e^{-\frac{aL}{2}}\right\}}{\left\{\dfrac{L}{2}-\dfrac{\sin(2w_n L)}{4w_n}\right\}}\right] \sum_{i=1}^{5} \frac{\gamma_i \mu_i e^{\mu_i t}}{\mu_i + \dfrac{a^2}{4b} + \dfrac{w_n^2}{b}}
\tag{20}
$$

Hence, the temperature field function is

$$
T = \sum_{i=1}^{5} \gamma_i e^{\mu_i t} +
$$

$$
\sum_{n=1}^{\infty} \sin(w_n x)\, e^{\frac{ax}{2}} K_n e^{-\frac{t}{b}\left(\frac{a^2}{4}+w_n^2\right)} + \sum_{n=1}^{\infty} \sin(w_n x)\, e^{\frac{ax}{2}} e^{-\frac{t}{b}\left(\frac{a^2}{4}+w_n^2\right)}
$$

$$
\left[\frac{\dfrac{2}{a^2+4w_n^2}\left\{2w_n - a\sin(w_n L)e^{-\frac{aL}{2}} - 2w_n\cos(w_n L)e^{-\frac{aL}{2}}\right\}}{\left\{\dfrac{L}{2}-\dfrac{\sin(2w_n L)}{4w_n}\right\}}\right] \sum_{i=1}^{5} \frac{\gamma_i \mu_i}{\mu_i + \dfrac{a^2}{4b} + \dfrac{w_n^2}{b}}
$$

$$
-\sum_{n=1}^{\infty} \sin(w_n x)\, e^{\frac{ax}{2}} \left[\frac{\dfrac{2}{a^2+4w_n^2}\left\{2w_n - a\sin(w_n L)e^{-\frac{aL}{2}} - 2w_n\cos(w_n L)e^{-\frac{aL}{2}}\right\}}{\left\{\dfrac{L}{2}-\dfrac{\sin(2w_n L)}{4w_n}\right\}}\right] \sum_{i=1}^{5} \frac{\gamma_i \mu_i e^{\mu_i t}}{\mu_i + \dfrac{a^2}{4b} + \dfrac{w_n^2}{b}}
\tag{21}
$$

Expanding back K_n, a, b and taking $x=L$, the exit temperature function becomes

$$T_{exit} = \sum_{i=1}^{5} \gamma_i e^{\mu_i t}$$

$$+ \sum_{n=1}^{\infty} \sin(w_n L) e^{\frac{\dot{m}_a c_{pa} L}{2 k_s BH}} e^{-\frac{k_s t}{\rho_s c_{ps}}\left(\frac{\dot{m}_a^2 c_{pa}^2}{4 k_s^2 B^2 H^2} + w_n^2\right)} \left(T_0 - \sum_{i=1}^{5} \gamma_i\right) \left[\frac{\dfrac{2}{\dfrac{\dot{m}_a^2 c_{pa}^2}{k_s^2 B^2 H^2} + 4 w_n^2}\left\{2 w_n - \dfrac{\dot{m}_a c_{pa}}{k_s BH} \sin(w_n L) e^{-\frac{\dot{m}_a c_{pa} L}{2 k_s BH}} - 2 w_n \cos(w_n L) e^{-\frac{\dot{m}_a c_{pa} L}{2 k_s BH}}\right\}}{\dfrac{L}{2} - \dfrac{\sin(2 w_n L)}{4 w_n}} \right]$$

$$+ \sum_{n=1}^{\infty} \sin(w_n L) e^{\frac{\dot{m}_a c_{pa} L}{2 k_s BH}} e^{-\frac{k_s t}{\rho_s c_{ps}}\left(\frac{\dot{m}_a^2 c_{pa}^2}{4 k_s^2 B^2 H^2} + w_n^2\right)}$$

$$\left[\frac{\dfrac{2}{\dfrac{\dot{m}_a^2 c_{pa}^2}{k_s^2 B^2 H^2} + 4 w_n^2}\left\{2 w_n - \dfrac{\dot{m}_a c_{pa}}{k_s BH} \sin(w_n L) e^{-\frac{\dot{m}_a c_{pa} L}{2 k_s BH}} - 2 w_n \cos(w_n L) e^{-\frac{\dot{m}_a c_{pa} L}{2 k_s BH}}\right\}}{\left\{\dfrac{L}{2} - \dfrac{\sin(2 w_n L)}{4 w_n}\right\}} \right] \sum_{i=1}^{5} \frac{\gamma_i \mu_i}{\mu_i + \dfrac{\dot{m}_a^2 c_{pa}^2}{4 k_s B^2 H^2 \rho_s c_{ps}} + \dfrac{k_s w_n^2}{\rho_s c_{ps}}}$$

$$- \sum_{n=1}^{\infty} \sin(w_n L) e^{\frac{\dot{m}_a c_{pa} L}{2 k_s BH}} \left[\frac{\dfrac{2}{\dfrac{\dot{m}_a^2 c_{pa}^2}{k_s^2 B^2 H^2} + 4 w_n^2}\left\{2 w_n - \dfrac{\dot{m}_a c_{pa}}{k_s BH} \sin(w_n L) e^{-\frac{\dot{m}_a c_{pa} L}{2 k_s BH}} - 2 w_n \cos(w_n L) e^{-\frac{\dot{m}_a c_{pa} L}{2 k_s BH}}\right\}}{\left\{\dfrac{L}{2} - \dfrac{\sin(2 w_n L)}{4 w_n}\right\}} \right] \sum_{i=1}^{5} \frac{\gamma_i \mu_i e^{\mu_i t}}{\mu_i + \dfrac{\dot{m}_a^2 c_{pa}^2}{4 k_s B^2 H^2 \rho_s c_{ps}} + \dfrac{k_s w_n^2}{\rho_s c_{ps}}} \tag{22}$$

Question 22

The vessel shown contains two compartments, having n moles of an ideal gas each, at temperatures T_{1i} and T_{2i} $(T_{1i} > T_{2i})$. The piston (of mass m) is movable and is held in the mid position initially. At $t=0$, it is released. Find piston velocity as a function of its displacement. The piston is adiabatic and so are all the compartment walls. The universal gas constant is R and the adiabatic exponent of the gas (same gas in both compartments) is γ

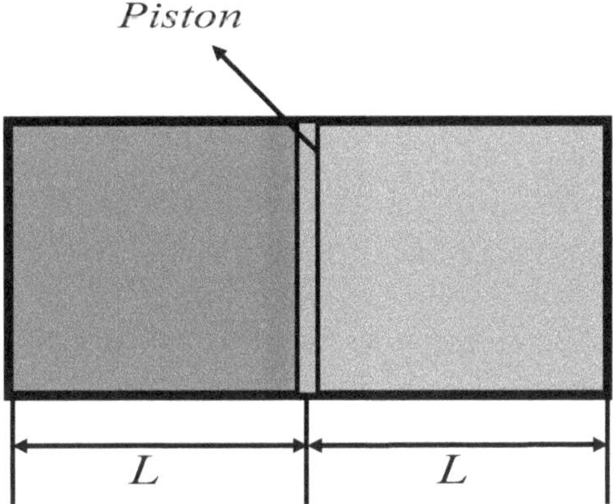

Solution

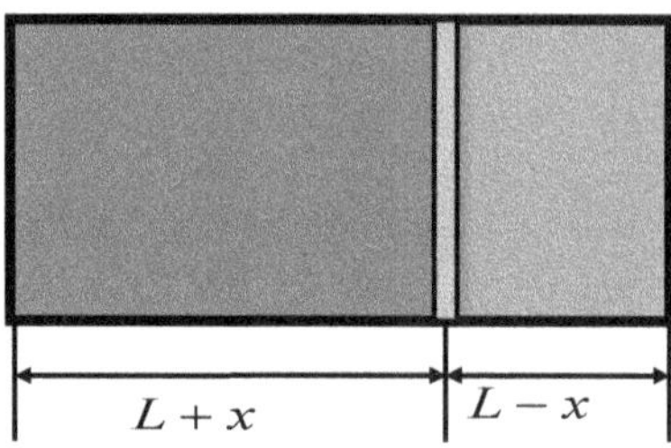

Applying Newton's second law of motion on the piston at a general location gives

$$P_1 A - P_2 A = m(acc) \tag{1}$$

Using ideal gas equation for each compartment

$$\left[\frac{nRT_1}{A(L+x)}\right]A - \left[\frac{nRT_2}{A(L-x)}\right]A = mv\frac{dv}{dx}$$

$$\Rightarrow v\frac{dv}{dx} = \frac{nR}{m}\left(\frac{T_1}{L+x} - \frac{T_2}{L-x}\right) \tag{2}$$

Because of perfect thermal insulation, the process is adiabatic for the gas contained in each compartment. Using the relation between temperature and volume of an ideal gas for an adiabatic process,

$$T_{1i}(AL)^{\gamma-1} = T_1\{A(L+x)\}^{\gamma-1}$$

$$\Rightarrow T_1 = \frac{T_{1i}L^{\gamma-1}}{(L+x)^{\gamma-1}} \tag{3}$$

$$T_{2i}(AL)^{\gamma-1} = T_2\{A(L-x)\}^{\gamma-1}$$

$$\Rightarrow T_2 = \frac{T_{2i}L^{\gamma-1}}{(L-x)^{\gamma-1}} \tag{4}$$

Feeding equations 3 and 4 in equation 2

$$v\frac{dv}{dx} = \frac{nRL^{\gamma-1}}{m}\left(\frac{T_{1i}}{(L+x)^{\gamma}} - \frac{T_{2i}}{(L-x)^{\gamma}}\right)$$

(5)

Separating the variables and integrating both sides

$$\int_{0}^{v} vdv = \frac{nRL^{\gamma-1}}{m}\int_{0}^{x}\left(\frac{T_{1i}}{(L+x)^{\gamma}} - \frac{T_{2i}}{(L-x)^{\gamma}}\right)dx$$

(6)

$$\frac{v^2}{2} = \frac{nRL^{\gamma-1}}{m}\left[\frac{T_{1i}(L+x)^{1-\gamma}}{1-\gamma} + \frac{T_{2i}(L-x)^{1-\gamma}}{1-\gamma}\right]_{0}^{x}$$

$$\Rightarrow v = \sqrt{\frac{2nRL^{\gamma-1}\left[T_{1i}(L+x)^{1-\gamma} + T_{2i}(L-x)^{1-\gamma} - T_{1i}L^{1-\gamma} - T_{2i}L^{1-\gamma}\right]}{m(1-\gamma)}}$$

(7)

Question 23

❖ ❖ ❖

The vessel shown contains two compartments, having n moles of an ideal gas each, at temperatures T_{1i} and T_{2i} $(T_{1i} > T_{2i})$. The vessel is perfectly insulated from the surroundings. The piston is fixed and is diathermic. However, we have two movable, perfect insulation layers on either side of it. A $t=0$, the insulation layers are removed. Find the compartment temperatures as functions of time. The overall heat transfer coefficient between the two gas masses is U. The cross-section area of the piston is A. The universal gas constant is R and the adiabatic exponent of the gas (same gas in both compartments) is γ

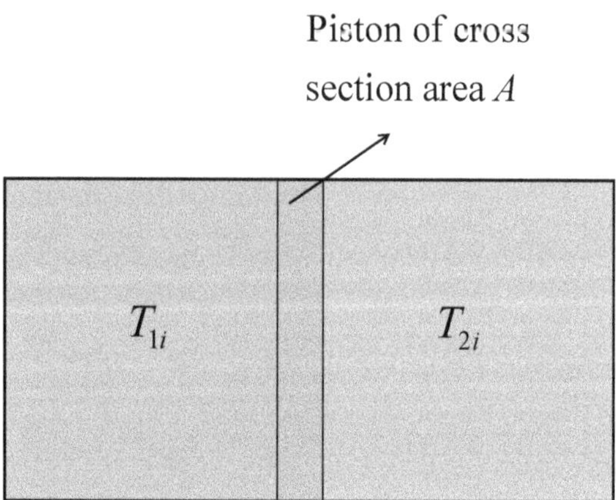

Solution

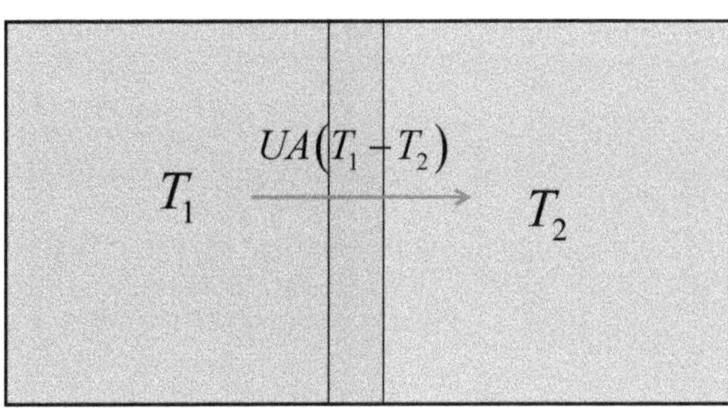

Applying law of energy conservation for the two gas masses at a general time t

$$\frac{d(nC_vT_1)}{dt} = -UA(T_1 - T_2) \tag{1}$$

$$\frac{d(nC_vT_2)}{dt} = UA(T_1 - T_2) \tag{2}$$

Equations 1 and 2 can be written as

$$\frac{dT_1}{dt} = -a(T_1 - T_2) \tag{3}$$

$$\frac{dT_2}{dt} = a(T_1 - T_2) \tag{4}$$

where $a = \dfrac{UA}{nC_v} = \dfrac{UA(\gamma - 1)}{nR}$. Feeding T_2 from Equation 3 in 4

$$\frac{d^2T_1}{dt^2} + 2a\frac{dT_1}{dt} = 0 \tag{5}$$

Separating the variables and integrating both sides

$$\frac{d}{dt}\left(\frac{dT_1}{dt}\right) + 2a\frac{dT_1}{dt} = 0$$

$$\Rightarrow d\left(\frac{dT_1}{dt}\right) + 2a \cdot dT_1 = 0$$

$$\Rightarrow \int d\left(\frac{dT_1}{dt}\right) + 2a\int dT_1 = C_1$$

$$\Rightarrow \frac{dT_1}{dt} + 2aT_1 = C_1 \tag{6}$$

where C_1 is an arbitrary integration constant. Separating the variables and integrating both sides

$$\int_{T_{1i}}^{T_1} \frac{dT_1}{C_1 - 2aT_1} = \int_0^t dt$$

$$\Rightarrow \left[\frac{\ln|C_1 - 2aT_1|}{-2a}\right]_{T_{1i}}^{T_1} = t \tag{7}$$

Simplifying

$$\frac{C_1 - 2aT_1}{C_1 - 2aT_{1i}} = e^{-2at}$$

$$\Rightarrow C_1 - 2aT_1 = \left(C_1 - 2aT_{1i}\right)e^{-2at}$$

$$\Rightarrow T_1 = \left\{\frac{C_1}{2a}\right\} - \left\{\left(\frac{C_1}{2a} - T_{1i}\right)e^{-2at}\right\} \tag{8}$$

Feeding Equation 8 in 3

$$T_2 = \left\{\frac{C_1}{2a}\right\} + \left\{\left(\frac{C_1}{2a} - T_{1i}\right)e^{-2at}\right\} \tag{9}$$

The constant C_1 is to be found from the initial condition: $(T_2)_{t=0}=T_{2i}$

$$C_1 = a\left(T_{1i} + T_{2i}\right) \tag{10}$$

Feeding equation 10 in 8 and 9, and expanding back a

$$T_1 = \left\{\frac{T_{1i} + T_{2i}}{2}\right\} - \left\{\left(\frac{T_{2i} - T_{1i}}{2}\right)e^{-\frac{2UA(\gamma-1)t}{nR}}\right\} \tag{11}$$

$$T_2 = \left\{\frac{T_{1i} + T_{2i}}{2}\right\} + \left\{\left(\frac{T_{2i} - T_{1i}}{2}\right)e^{-\frac{2UA(\gamma-1)t}{nR}}\right\} \tag{12}$$

Question 24

The Redlich Kwong equation of state for a gas is given by $P = \dfrac{nRT}{V - nb} - \dfrac{an^2}{V(V + nb)\sqrt{T}}$ where P is gas pressure, T is gas temperature, V is gas volume, n its number of moles, a and b are constants depending on gas nature, R is the universal gas constant. Derive an expression relating temperature and volume for a gas obeying this equation of state when it undergoes an adiabatic process. Take $b=0$, and molar specific heat capacity of the gas at constant volume is C_v

Solution

The differential form of 1st law of thermodynamics is

$$dQ = dU + dW \tag{1}$$

Using the expressions for differential changes in internal energy and work done

$$dQ = nC_v \cdot dT - \left\{ P - T\left(\frac{\partial P}{\partial T}\right)_V \right\} dV + P \cdot dV \tag{2}$$

For an adiabatic process, $dQ=0$.

$$nC_v \cdot dT + T\left(\frac{\partial P}{\partial T}\right)_V dV = 0 \tag{3}$$

As per the given equation of state, the partial derivative appearing in equation 3 is

$$\left(\frac{\partial P}{\partial T}\right)_V = \frac{nR}{V} + \frac{an^2}{2V^2 T\sqrt{T}} \tag{4}$$

Feeding equation 4 in 3

$$nC_v \cdot dT + \left(\frac{nRT}{V} + \frac{an^2}{2V^2\sqrt{T}}\right)dV = 0 \tag{5}$$

Rearranging

$$\frac{dT}{dV} + \frac{RT}{C_v V} = -\frac{an}{2V^2 C_v \sqrt{T}} \tag{6}$$

Multiplying throughout by $\sqrt{T}$

$$\frac{\sqrt{T}\,dT}{dV} + \frac{RT^{\frac{3}{2}}}{C_v V} = -\frac{an}{2V^2 C_v} \tag{7}$$

Let $T^{\frac{3}{2}} = p \Rightarrow \frac{3}{2}\sqrt{T}\,dT = dp$

$$\frac{2}{3}\frac{dp}{dV} + \frac{Rp}{C_v V} = -\frac{an}{2V^2 C_v}$$

$$\Rightarrow \frac{dp}{dV} + \frac{3Rp}{2C_v V} = -\frac{3an}{4V^2 C_v} \tag{8}$$

Equation 8 is a first order linear ordinary differential equation which can be solved by the integrating factor method. The general solution is

$$p(\text{I.F}) = \int -\frac{3an(\text{I.F})}{4V^2 C_v} \cdot dV + C_0 \tag{9}$$

Where C_0 is an arbitrary integration constant and I.F is the relevant integrating factor given by

$$\text{I.F} = e^{\int \frac{3R \cdot dV}{2C_v V}} = e^{\frac{3R}{2C_v} \int \frac{dV}{V}} = e^{\frac{3R}{2C_v} \ln|V|} = e^{\ln|V|\frac{3R}{2C_v}} = V^{\frac{3R}{2C_v}} \tag{10}$$

Feeding equation 10 in 9

$$pV^{\frac{3R}{2C_v}} = \left(\int -\frac{3anV^{\frac{3R}{2C_v}}}{4V^2 C_v} \cdot dV \right) + C_0$$

$$\Rightarrow pV^{\frac{3R}{2C_v}} = \left(-\frac{3an}{4C_v} \int V^{\frac{3R}{2C_v}-2} \cdot dV \right) + C_0$$

$$\Rightarrow pV^{\frac{3R}{2C_v}} = \left\{ -\frac{3anV^{\frac{3R}{2C_v}-1}}{4C_v \left(\frac{3R}{2C_v} - 1 \right)} \right\} + C_0$$

$$\Rightarrow p = -\frac{3an}{2V(3R - 2C_v)} + C_0 V^{-\frac{3R}{2C_v}}$$

$$\Rightarrow T^{\frac{3}{2}} = -\frac{3an}{2V(3R - 2C_v)} + C_0 V^{-\frac{3R}{2C_v}} \tag{11}$$

Rearranging

$$T^{\frac{3}{2}}V^{\frac{3R}{2C_v}} + \frac{3anV^{\frac{3R}{2C_v}-1}}{2\left(3R-2C_v\right)} = \text{constant}$$

$$(12)$$

Question 25

The Berthelot equation of state for a gas is given by $P = \dfrac{nRT}{V - nb} - \dfrac{an^2}{TV^2}$ where P is gas pressure, T is gas temperature, V is gas volume, n its number of moles, a and b are constants depending on gas nature, R is the universal gas constant. A gas obeying this equation of state expands isothermally from volume V_1 to V_2. The work done by the gas in this process is W. At the end of this process, the gas is supplied heat at constant volume by a heater. The rate of heat provided by the heater decays exponentially with time as per the law: $\dot{Q} = \dot{Q}_0 e^{-kt}$. Find the final temperature of the gas if heat is supplied for a time t_0. The molar specific heat capacity of the gas at constant volume is C_v

Solution

Work done by a gas undergoing any process is given by $\displaystyle\int_{V_1}^{V_2} P \cdot dV$.

Pressure needs to be expressed as a function of V alone to evaluate this integral. Here, this can be done by feeding P from the given equation of state.

$$W = \int_{V_1}^{V_2} \left[\frac{nRT}{V - nb} - \frac{an^2}{TV^2} \right] dV \tag{1}$$

Simplifying

$$W = \left(nRT \int_{V_1}^{V_2} \frac{dV}{V - nb} \right) - \left(\frac{an^2}{T} \int_{V_1}^{V_2} \frac{dV}{V^2} \right) \tag{2}$$

$$W = nRT \ln \left| \frac{V_2 - nb}{V_1 - nb} \right| + \frac{an^2}{T} \left(\frac{1}{V_2} - \frac{1}{V_1} \right)$$

$$\Rightarrow W = nRT \ln \left| \frac{V_2 - nb}{V_1 - nb} \right| + \frac{an^2}{T} \left(\frac{V_1 - V_2}{V_1 V_2} \right) \tag{3}$$

Multiplying throughout by T

$$nR \ln \left| \frac{V_2 - nb}{V_1 - nb} \right| T^2 - WT + an^2 \left(\frac{V_1 - V_2}{V_1 V_2} \right) = 0 \tag{4}$$

Equation 4 is a quadratic in T. Therefore,

$$T = \frac{W \pm \sqrt{W^2 + 4an^3 R \left(\frac{V_2 - V_1}{V_1 V_2} \right) \ln \left| \frac{V_2 - nb}{V_1 - nb} \right|}}{2nR \ln \left| \frac{V_2 - nb}{V_1 - nb} \right|} \tag{5}$$

The quantity inside the square root symbol in equation 5 is greater than W^2. Therefore, magnitude of its square root will be greater than W. Thus, taking minus sign in equation 5 will result in negative temperature which isn't physically meaningful. Thus, only the positive sign is considered

$$T = \frac{W + \sqrt{W^2 + 4an^3 R \left(\dfrac{V_2 - V_1}{V_1 V_2} \right) \ln \left| \dfrac{V_2 - nb}{V_1 - nb} \right|}}{2nR \ln \left| \dfrac{V_2 - nb}{V_1 - nb} \right|} \tag{6}$$

Subsequent process is isochoric heating. The differential form of 1st law of thermodynamics is

$$dQ = dU + dW \tag{7}$$

Using the expressions for differential changes in internal energy and work done

$$dQ = nC_v \cdot dT - \left\{ P - T \left(\frac{\partial P}{\partial T} \right)_V \right\} dV + P \cdot dV \tag{8}$$

For an isochoric process, $dV = 0$.

$$dQ = nC_v \cdot dT \tag{9}$$

Differentiating both sides with respect to time

$$\frac{dQ}{dt} = nC_v \frac{dT}{dt} \tag{10}$$

Using given expression of heater's heat supply rate

$$nC_v \frac{dT}{dt} = \dot{Q}_0 e^{-kt} \tag{11}$$

Separating the variables and integrating both sides

$$\int_{T}^{T_f} dT = \int_{0}^{t_0} \frac{\dot{Q}_0 e^{-kt}}{nC_v} dt$$

$$\Rightarrow T_f = T + \frac{\dot{Q}_0 \left(1 - e^{-kt_0}\right)}{nkC_v}$$

(12)

Where T_f is the final temperature required to be found out. Feeding equation 6 in 12

$$T_f = \frac{W + \sqrt{W^2 + 4an^3 R \left(\dfrac{V_2 - V_1}{V_1 V_2}\right) \ln\left|\dfrac{V_2 - nb}{V_1 - nb}\right|}}{2nR \ln\left|\dfrac{V_2 - nb}{V_1 - nb}\right|} + \frac{\dot{Q}_0 \left(1 - e^{-kt_0}\right)}{nkC_v}$$

(13)

Question 26

❖ ❖ ❖

The Van der Waal equation of state for a gas is $\left(P + \dfrac{an^2}{V^2}\right)(V - nb) = nRT$ where P is gas pressure, T is gas temperature, V is gas volume, n its number of moles, a and b are constants depending on gas nature, R is the universal gas constant. 2 samples of a gas obeying this equation of state are kept inside 2 vessels of volumes V_1 and V_2. The number of moles in the two samples are n_1 and n_2 and they are respectively at temperatures T_1 and T_2. The molar specific heat capacity of the gas at constant volume is C_v. The 2 gas masses are allowed to mix by opening the valve shown in the figure. Find the final temperature and pressure of the combined gas mass. The entire apparatus is thermally insulated from the surroundings.

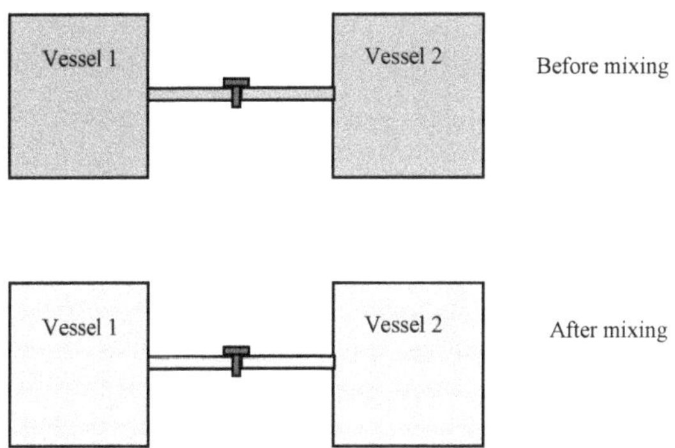

Solution

No heat is exchanged and no work is done by the composite system (of two containers) with the surroundings. Therefore, internal energy of this system shall remain constant. First, an expression for internal energy of a Van der Waal gas needs to be obtained and then using it, the fact that internal energy of the combined system is conserved needs to be mathematically expressed.

The general expression for differential change in internal energy is

$$dU = nC_v \cdot dT - \left\{ P - T\left(\frac{\partial P}{\partial T}\right)_V \right\} dV \tag{1}$$

Evaluating the partial derivative appearing in equation 1 using the given equation of state

$$dU = nC_v \cdot dT - \left\{ P - T\left(\frac{nR}{V - nb}\right) \right\} dV \tag{2}$$

Feeding P from given equation of state

$$dU = nC_v \cdot dT - \left\{ \frac{nRT}{V-nb} - \frac{an^2}{V^2} - \frac{nRT}{V-nb} \right\} dV$$

$$\Rightarrow dU = nC_v \cdot dT + \frac{an^2}{V^2} dV \tag{3}$$

Separating the variables and integrating both sides

$$\int dU = \int nC_v \cdot dT + \int \frac{an^2}{V^2} dV$$

$$\Rightarrow U = nC_v T - \frac{an^2}{V} + \text{constant} \tag{4}$$

Equation 4 provides an expression for internal energy of a Van der Waal gas. Equating internal energy of the combined system before and after mixing

$$n_1 C_v T_1 - \frac{an_1^2}{V_1} + n_2 C_v T_2 - \frac{an_2^2}{V_2} = (n_1 + n_2) C_v T_f - \frac{a(n_1 + n_2)^2}{V_1 + V_2} \tag{5}$$

Where T_f is the final, common temperature. [Since gases are allowed to mix, they will eventually attain a common final temperature owing to physical mixing]

$$T_f = \frac{n_1 C_v T_1 - \dfrac{an_1^2}{V_1} + n_2 C_v T_2 - \dfrac{an_2^2}{V_2} + \dfrac{a(n_1 + n_2)^2}{V_1 + V_2}}{(n_1 + n_2) C_v} \tag{6}$$

Feeding equation 6 in the mentioned equation of state yields an expression for common, final pressure

$$P_f = \frac{\left(n_1 + n_2\right)RT_f}{\left\{\left(V_1 + V_2\right)\right\} - \left\{\left(n_1 + n_2\right)b\right\}} - \frac{a\left(n_1 + n_2\right)^2}{\left(V_1 + V_2\right)^2}$$

$$\Rightarrow P_f = \frac{n_1 RT_1 - \dfrac{aRn_1^2}{V_1 C_v} + n_2 RT_2 - \dfrac{aRn_2^2}{V_2 C_v} + \dfrac{aR\left(n_1 + n_2\right)^2}{\left(V_1 + V_2\right)C_v}}{\left\{\left(V_1 + V_2\right)\right\} - \left\{\left(n_1 + n_2\right)b\right\}} - \frac{a\left(n_1 + n_2\right)^2}{\left(V_1 + V_2\right)^2} \tag{7}$$

Question 27

A solid sphere of radius R is rotated about its axis in a viscous fluid medium (of mass density ρ_f) at an angular velocity w_0 at time $t=0$. The drag force generated by the medium follows the law $\vec{F}_{drag} = -C_D\left(\frac{1}{2}\rho_f |\vec{v}|^2\right) A\hat{v}$ where C_D is the drag coefficient (consider it constant), A is area of the object in contact with the medium, and v is relative velocity of object with respect to the medium. Find energy dissipated upto time t if ρ_b is the mass density of the sphere. Consider the surrounding medium to be large and unaffected by the rotation.

Solution

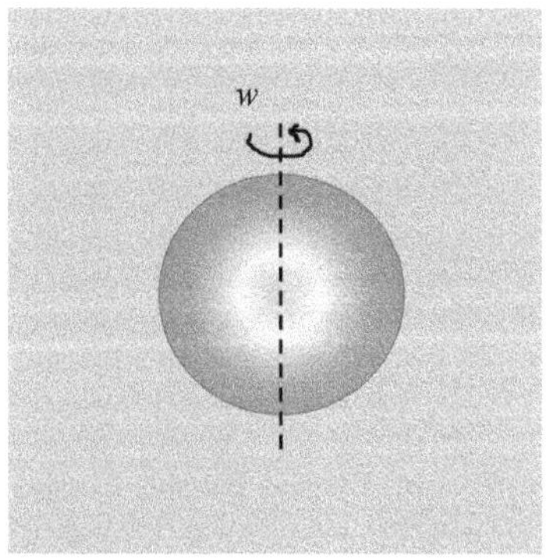

Every differential area of the sphere in contact with the surrounding air moves with respect to this air at a velocity wR. So the viscous force acting on any such little area will be

$$dF_{viscous} = -\frac{C_D \rho_f (dA)(wR)^2}{2} \tag{1}$$

Each such differential force vector acts at a distance R from the axis of rotation. Thus, the associated differential torque will be

$$d\tau_{viscous} = -\frac{C_D \rho_f (dA)(wR)^2 R}{2} \tag{2}$$

All such differential torques act in the same sense. Therefore, the total viscous torque acting on the sphere will be

$$\tau_{viscous} = \int d\tau_{viscous} = \int -\frac{C_D \rho_f (dA)(wR)^2 R}{2} = -\frac{C_D \rho_f w^2 R^3}{2} \int dA$$

$$\Rightarrow \tau_{viscous} = -\frac{C_D \rho_f w^2 R^3 (4\pi R^2)}{2} = -2\pi C_D \rho_f w^2 R^5 \tag{3}$$

Applying rotational analogue of newton's second law of motion on the sphere at a general time t

$$-2\pi C_D \rho_f w^2 R^5 = \frac{d\left(\frac{2}{5} MR^2 w\right)}{dt} = \frac{2}{5} MR^2 \frac{dw}{dt} \tag{4}$$

Where the term $\frac{2}{5} MR^2$ represents mass moment of inertia of a sphere about an axis passing through its centre, M being the sphere mass. Since M is not given in the question, it must be expressed in terms of provided parameters.

$$\frac{dw}{dt} = -\frac{5\pi C_D \rho_f w^2 R^3}{M} = -\frac{5\pi C_D \rho_f w^2 R^3}{\rho_b \frac{4}{3}\pi R^3} = -\frac{15 C_D \rho_f w^2}{4\rho_b}$$

(5)

Separating the variables and integrating both sides,

$$\int_{w_0}^{w} \frac{dw}{w^2} = \int_{0}^{t} \frac{-15 C_D \rho_f \cdot dt}{4\rho_b}$$

$$\Rightarrow \frac{1}{w_0} - \frac{1}{w} = -\frac{15 C_D \rho_f t}{4\rho_b}$$

(6)

Simplifying

$$\frac{1}{w_0} + \frac{15 C_D \rho_f t}{4\rho_b} = \frac{1}{w}$$

$$\Rightarrow \frac{4\rho_b + 15 C_D \rho_f w_0 t}{4\rho_b w_0} = \frac{1}{w}$$

$$\Rightarrow w = \frac{4\rho_b w_0}{4\rho_b + 15 C_D \rho_f w_0 t}$$

(7)

Energy dissipation rate is the rate of work done by the retarding viscous torque

$$P = \frac{dE}{dt} = \tau_{viscous}\, w$$

(8)

Feeding equations 3 and 7 in 8

$$\frac{dE}{dt} = -2\pi C_D \rho_f w^3 R^5 = -128\pi C_D \rho_f \rho_b^3 w_0^3 R^5 \left(4\rho_b + 15 C_D \rho_f w_0 t\right)^{-3}$$

(9)

Separating the variables and integrating both sides

$$\int dE = \int_0^t -128\pi C_D \rho_f \rho_b^3 w_0^3 R^5 \left(4\rho_b + 15 C_D \rho_f w_0 t\right)^{-3} \cdot dt$$

$$\Rightarrow E = -128\pi C_D \rho_f \rho_b^3 w_0^3 R^5 \left[\frac{\left(4\rho_b + 15 C_D \rho_f w_0 t\right)^{-2}}{-2\left(15 C_D \rho_f w_0\right)}\right]_0^t$$

$$\Rightarrow E = \frac{64\pi \rho_b^3 w_0^3 R^5}{15 w_0} \left[\frac{1}{\left(4\rho_b + 15 C_D \rho_f w_0 t\right)^2} - \frac{1}{\left(4\rho_b\right)^2}\right] \tag{10}$$

Question 28

❖❖❖

A disc of radius R, thickness δ rotates at an angular velocity w_0 about its axis. At $t=0$, it is placed on a rough surface. The friction coefficient for the disc surface pair is location dependent. At a radial distance r from the disc centre, this coefficient is given by $\mu = ae^{-br}$ where a, b are positive constants. Find the instantaneous heat generation rate. The mass density of disc is ρ_b and acceleration due to gravity on the Earth's surface is g.

Solution

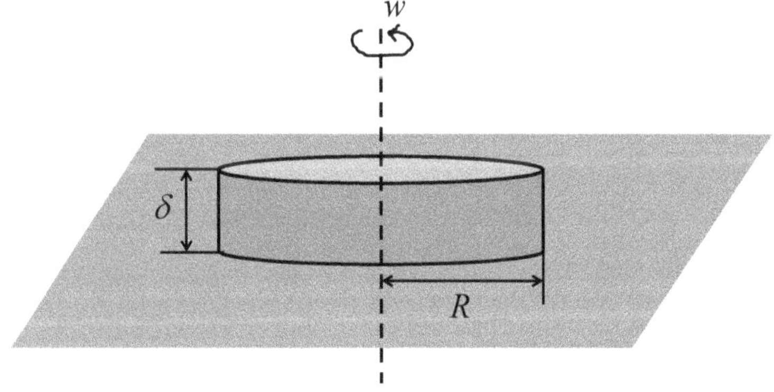

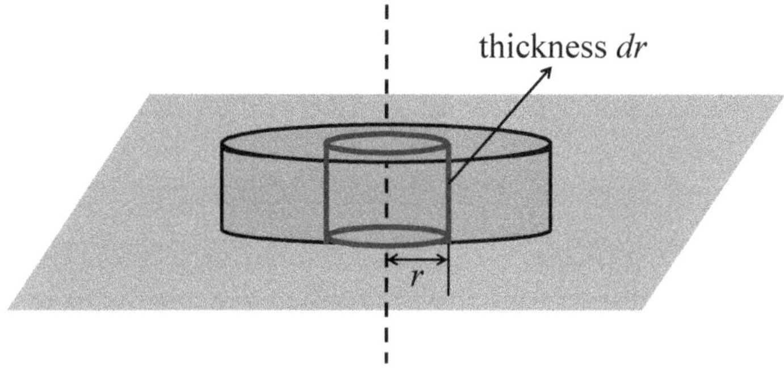

The differential torque acting on a differential ring as shown is the product of frictional force acting on it and its radius. The differential friction force will be friction coefficient times the differential normal reaction acting on the ring. And the differential normal reaction will be the weight of this ring [for vertical equilibrium]

$$d\tau = dF \cdot r = -\mu dN \cdot r = -\mu dm \cdot g \cdot r \tag{1}$$

Mass of the ring will be the product of its volume and density of the disc material

$$dm = \rho_b \left(2\pi r \cdot dr\right) \delta \tag{2}$$

Feeding equation 2 in 1 and using the mentioned function of friction coefficient

$$d\tau = -2\pi \rho_b \delta g a r^2 e^{-br} \cdot dr \tag{3}$$

All differential retarding torques on all such differential rings will act in the same sense. Thus, the total retarding torque will be

$$\tau = \int d\tau = -2\pi\rho_b \delta ga \int_0^R r^2 e^{-br} \cdot dr \tag{4}$$

To proceed further, the integral appearing in equation 4 needs to be evaluated. This can be done using integration by parts. For convenience, let the corresponding indefinite integral be termed "Integral"

$$\text{Integral} = r^2 \left(\frac{e^{-br}}{-b} \right) - \int 2r \left(\frac{e^{-br}}{-b} \right) \cdot dr$$

$$\Rightarrow \text{Integral} = -\frac{r^2 e^{-br}}{b} + \frac{2}{b}\int re^{-br} \cdot dr \tag{5}$$

Using integration by parts on the integral appearing in equation 5,

$$\text{Integral} = -\frac{r^2 e^{-br}}{b} + \frac{2}{b}\left[r\left(\frac{e^{-br}}{-b} \right) - \int \left(\frac{e^{-br}}{-b} \right) \cdot dr \right]$$

$$\Rightarrow \text{Integral} = -\frac{r^2 e^{-br}}{b} - \frac{2re^{-br}}{b^2} - \frac{2e^{-br}}{b^3}$$

$$\Rightarrow \text{Integral} = -\frac{e^{-br}}{b}\left(r^2 + \frac{2r}{b} + \frac{2}{b^2} \right) \tag{6}$$

Feeding the relevant limits in the above integral and then substituting the definite integral so obtained in equation 4

$$\tau = -2\pi\rho_b \delta ga \left[\frac{2}{b^3} - \frac{e^{-bR}\left(R^2 + \frac{2R}{b} + \frac{2}{b^2} \right)}{b} \right] \tag{7}$$

Applying rotational analogue of newton's second law of motion on the disc at a general time t

$$\tau = \frac{d\left(\frac{1}{2}MR^2 w\right)}{dt} = \frac{1}{2}MR^2 \frac{dw}{dt} \tag{8}$$

Where the term $\frac{1}{2}MR^2$ represents mass moment of inertia of a disc about an axis passing through its centre and perpendicular to its plane, M being the disc mass. Since M is not given in the question, it must be expressed in terms of provided parameters.

$$\tau = \frac{1}{2}\left(\rho_b \cdot \pi R^2 \delta\right) R^2 \frac{dw}{dt} \tag{9}$$

Feeding equation 7 in 9

$$\frac{dw}{dt} = -\frac{4ga}{R^4}\left[\frac{2}{b^3} - \frac{e^{-bR}\left(R^2 + \frac{2R}{b} + \frac{2}{b^2}\right)}{b}\right] \tag{10}$$

Separating the variables and integrating both sides

$$\int_{w_0}^{w} dw = \int_{0}^{t} -\frac{4ga}{R^4}\left[\frac{2}{b^3} - \frac{e^{-bR}\left(R^2 + \frac{2R}{b} + \frac{2}{b^2}\right)}{b}\right] dt$$

$$\Rightarrow w = w_0 - \frac{4gat}{R^4}\left[\frac{2}{b^3} - \frac{e^{-bR}\left(R^2 + \frac{2R}{b} + \frac{2}{b^2}\right)}{b}\right] \tag{11}$$

The instantaneous heat generation rate will be the instantaneous rate of energy dissipation by the frictional force.

$$\text{Heat generation rate}=|\tau w| \tag{12}$$

Feeding equations 7, 11 in 12

Heat generation rate

$$=2\pi\rho_b\delta ga\left\{\frac{2}{b^3}-\frac{e^{-bR}\left(R^2+\dfrac{2R}{b}+\dfrac{2}{b^2}\right)}{b}\right\}\left[w_0-\frac{4gat\left\{\dfrac{2}{b^3}-\dfrac{e^{-bR}\left(R^2+\dfrac{2R}{b}+\dfrac{2}{b^2}\right)}{b}\right\}}{R^4}\right] \tag{13}$$

Question 29

❖ ❖ ❖

A solid cone of radius R, height H is rotating at an angular velocity w_0 about its axis. At $t=0$, it is placed on a rough surface wherein the friction coefficient for the cone surface pair is μ. Air resistance is operative as per the law $\vec{F}_{drag} = -C_D\left(\dfrac{1}{2}\rho_f |\vec{v}|^2\right)A\hat{v}$ where C_D is the drag coefficient (consider it constant), ρ_f is density of air, A is area of the object in contact with air, and v is relative velocity of object with respect to air. Find instantaneous heat generation rate if mass density of the cone material is ρ_b. The acceleration due to gravity on the Earth's surface is g. Consider the surrounding air mass to be large and unaffected by the rotation.

Solution

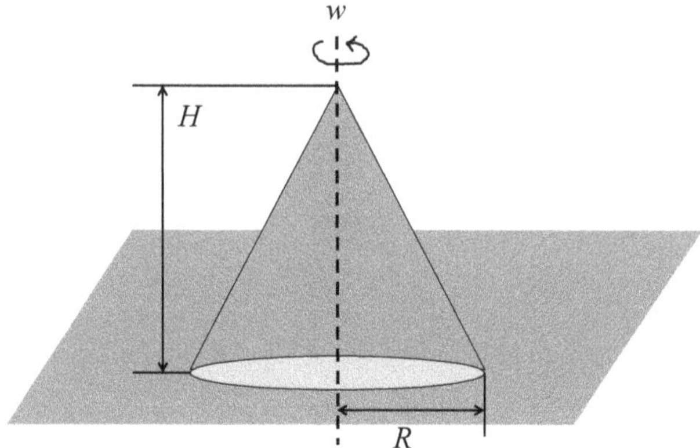

Both surface and viscous friction will be active in this problem. The former shall act on the base of the cone while the latter shall operate on its curved surface.

Evaluation of Surface Friction Torque:

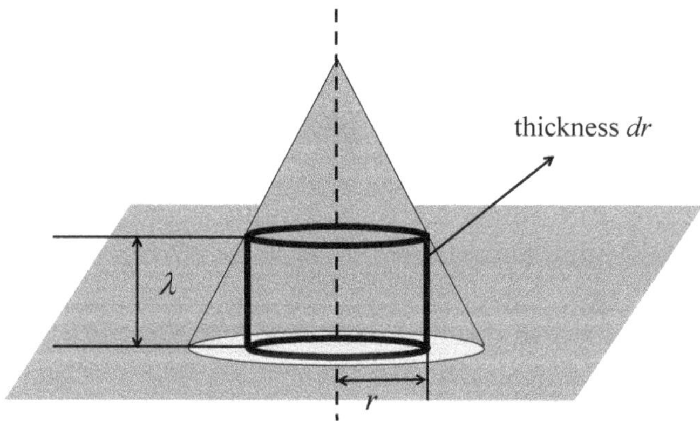

The differential surface friction torque acting on a differential cylinder as shown is the product of frictional force acting on it and its radius. The differential friction force will be friction coefficient times the differential normal reaction acting on the cylinder. And the differential normal reaction will be the weight of this cylinder [for vertical equilibrium]

$$d\tau_s = dF_s \cdot r = -\mu dN \cdot r = -\mu dm \cdot g \cdot r \tag{1}$$

Mass of the differential cylinder will be the product of its volume and density of the cone material

$$dm = \rho_b \left(2\pi r \cdot dr \cdot \lambda \right) = \left(2\pi \rho_b \lambda \right) r \cdot dr \tag{2}$$

Where λ is the height of the cylinder. It can be expressed in terms of mentioned parameters using geometry.

$$\tan\left(\text{semi vertical angle of cone}\right) = \frac{r}{H-\lambda} = \frac{R}{H}$$

$$\Rightarrow \lambda = H\left(1 - \frac{r}{R} \right) \tag{3}$$

Feeding equations 2 and 3 in 1

$$d\tau_s = -2\pi \rho_b \mu g H \left(r^2 - \frac{r^3}{R} \right) \cdot dr \tag{4}$$

Net surface friction torque is

$$\tau_s = \int d\tau_s = \int_0^R -2\pi \rho_b \mu g H \left(r^2 - \frac{r^3}{R} \right) \cdot dr$$

$$\Rightarrow \tau_s = -\frac{\pi \rho_b \mu g H R^3}{6} \tag{5}$$

Evaluation of Viscous Friction Torque:

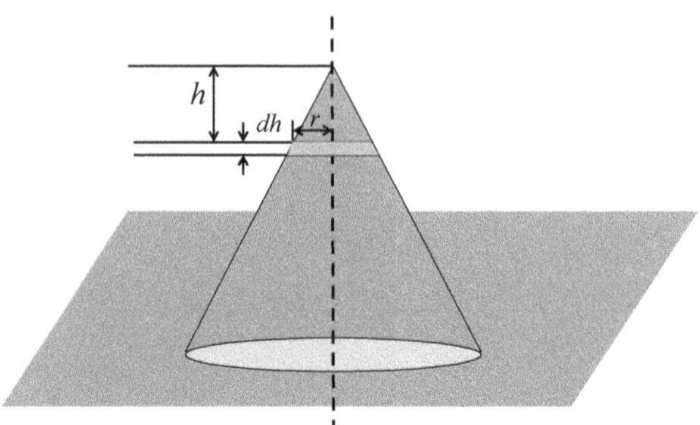

Every little area element on the differential cone shown, which is in contact with the surrounding air moves with respect to this air at a velocity rw. Further, each such differential force vector acts at a distance r from the axis of rotation. Thus, the associated differential torque will be

$$d\tau_v = dF_v \cdot r = -C_D \left(\frac{\rho_f r^2 w^2}{2} \right) (\pi r \cdot dl) \cdot r = -\frac{C_D \pi \rho_f r^4 w^2 dl}{2} \tag{6}$$

Using geometry, r can be expressed in terms of h

$$\tan\left(\text{semi vertical angle of cone}\right) = \frac{r}{h} = \frac{R}{H}$$

$$\Rightarrow r = \frac{Rh}{H} \tag{7}$$

Also, the differential of slant height can be expressed in terms of h as

$$dl = \sqrt{(dr)^2 + (dh)^2} = \sqrt{\frac{(dh)^2 R^2}{H^2} + (dh)^2} \quad \left[\text{Since } r = \frac{Rh}{H} \Rightarrow dr = \frac{R}{H} dh\right]$$

$$\Rightarrow dl = \frac{dh}{H}\sqrt{H^2 + R^2} \tag{8}$$

Feeding equations 7 and 8 in 6

$$d\tau_v = -\frac{C_D \pi \rho_f w^2 R^4 h^4 \sqrt{H^2 + R^2}}{2H^5} dh \tag{9}$$

Net viscous friction torque is

$$\tau_v = \int d\tau_v = \int_0^H -\frac{C_D \pi \rho_f w^2 R^4 h^4 \sqrt{H^2 + R^2}}{2H^5} dh$$

$$\Rightarrow \tau_v = -\frac{C_D \pi \rho_f w^2 R^4 \sqrt{H^2 + R^2}}{10} \tag{10}$$

Total friction torque is the sum of net surface and net viscous friction torque. Applying rotational analogue of newton's second law of motion on the cone at a general time t

$$\tau_{net} = \tau_s + \tau_v = \frac{d\left(\frac{3}{10}MR^2 w\right)}{dt}$$

$$\Rightarrow \tau_s + \tau_v = \frac{3}{10}MR^2 \frac{dw}{dt} \tag{11}$$

Where the term $\frac{3}{10}MR^2$ represents mass moment of inertia of a solid cone about its axis, M being the cone mass. Since M is not given in the question, it must be expressed in terms of provided parameters.

$$\tau_s + \tau_v = \frac{3}{10}\rho_b \left(\frac{\pi R^2 H}{3}\right) R^2 \frac{dw}{dt}$$

$$\Rightarrow \tau_s + \tau_v = \frac{\rho_b \pi R^4 H}{10}\frac{dw}{dt} \tag{12}$$

Feeding equations 5 and 10 in 12

$$\frac{\rho_b \pi R^4 H}{10}\frac{dw}{dt} = -\frac{\pi \rho_b \mu g H R^3}{6} - \frac{C_D \pi \rho_f w^2 R^4 \sqrt{H^2 + R^2}}{10}$$

$$\Rightarrow \frac{dw}{dt} = -a_0 - b_0 w^2 \tag{13}$$

Where $a_0 = \dfrac{5\mu g}{3R}, b_0 = \dfrac{C_D \rho_f \sqrt{H^2 + R^2}}{\rho_b H}$

Separating the variables and integrating both sides

$$\int_{w_0}^{w} \frac{dw}{a_0 + b_0 w^2} = \int_0^t -dt \tag{14}$$

Simplifying

$$\int_{w_0}^{w} \frac{dw}{b_0 \left(w^2 + \dfrac{a_0}{b_0}\right)} = \int_0^t -dt \tag{15}$$

$$\int_{w_0}^{w} \frac{dw}{w^2 + \left(\sqrt{\dfrac{a_0}{b_0}}\right)^2} = -b_0 t \tag{16}$$

Using the standard result $\displaystyle \int \frac{dZ}{Z^2 + S^2} = \frac{1}{S} \tan^{-1}\left(\frac{Z}{S}\right)$

$$\sqrt{\frac{b_0}{a_0}}\left[\tan^{-1}\left(w\sqrt{\frac{b_0}{a_0}}\right) - \tan^{-1}\left(w_0\sqrt{\frac{b_0}{a_0}}\right)\right] = -b_0 t$$

$$\Rightarrow \tan^{-1}\left(w\sqrt{\frac{b_0}{a_0}}\right) - \tan^{-1}\left(w_0\sqrt{\frac{b_0}{a_0}}\right) = -t\sqrt{a_0 b_0}$$

$$\Rightarrow \tan^{-1}\left(w\sqrt{\frac{b_0}{a_0}}\right) = \tan^{-1}\left(w_0\sqrt{\frac{b_0}{a_0}}\right) - t\sqrt{a_0 b_0}$$

$$\Rightarrow w = \sqrt{\frac{a_0}{b_0}} \tan\left[\tan^{-1}\left(w_0\sqrt{\frac{b_0}{a_0}}\right) - t\sqrt{a_0 b_0}\right] \tag{17}$$

The instantaneous heat generation rate will be the instantaneous rate of energy dissipation by the overall friction force

$$\text{Heat generation rate} = \left|\tau_{net}\, w\right| = \left|\left(\tau_s + \tau_v\right) w\right| \tag{18}$$

Feeding equations 5, 10, 17 in 18

Heat generation rate

$$= \frac{\pi \rho_b \mu g H R^3}{6} \sqrt{\frac{a_0}{b_0}} \tan\left[\tan^{-1}\left(w_0\sqrt{\frac{b_0}{a_0}}\right) - t\sqrt{a_0 b_0}\right] +$$

$$\frac{C_D \pi \rho_f R^4 \sqrt{H^2 + R^2}}{10} \frac{a_0}{b_0} \sqrt{\frac{a_0}{b_0}} \tan^3\left[\tan^{-1}\left(w_0\sqrt{\frac{b_0}{a_0}}\right) - t\sqrt{a_0 b_0}\right] \tag{19}$$

Expanding back a_0, b_0

Heat generation rate

$$= \frac{\pi \rho_b \mu g H R^3}{6} \sqrt{\frac{5\mu g \rho_b H}{3RC_D \rho_f \sqrt{H^2 + R^2}}} \tan\left[\tan^{-1}\left(w_0 \sqrt{\frac{3RC_D \rho_f \sqrt{H^2 + R^2}}{5\mu g \rho_b H}}\right) - t\sqrt{\frac{5\mu g C_D \rho_f \sqrt{H^2 + R^2}}{3R\rho_b H}}\right] +$$

$$\frac{C_D \pi \rho_f R^4 \sqrt{H^2 + R^2}}{10}\left(\frac{5\mu g \rho_b H}{3RC_D \rho_f \sqrt{H^2 + R^2}}\right)^{\frac{3}{2}} \tan^3$$

$$\left[\tan^{-1}\left(w_0 \sqrt{\frac{3RC_D \rho_f \sqrt{H^2 + R^2}}{5\mu g \rho_b H}}\right) - t\sqrt{\frac{5\mu g C_D \rho_f \sqrt{H^2 + R^2}}{3R\rho_b H}}\right] \tag{20}$$

Question 30

❖ ❖ ❖

A parallel flow heat exchanger is shown below. The mass flow rates of hot and cold fluids are: $\dot{m}_h, \dot{m}_c$. Their specific heat capacities are c_{ph}, c_{pc}. The overall heat transfer coefficient between the two fluids is U, and between cold fluid and ambience is U_l. The inlet temperatures of hot and cold fluids are T_{hi}, T_{ci}. Find outlet temperature of the cold fluid under steady state. Neglect conduction within each fluid and take ambient temperature to be T_a (consider it constant because of large size of surroundings). Neglect heat conduction, viscous effects and pressure gradients within each fluid.

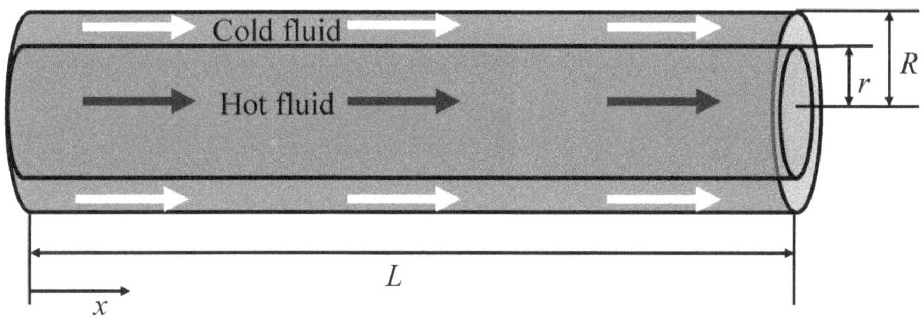

Solution

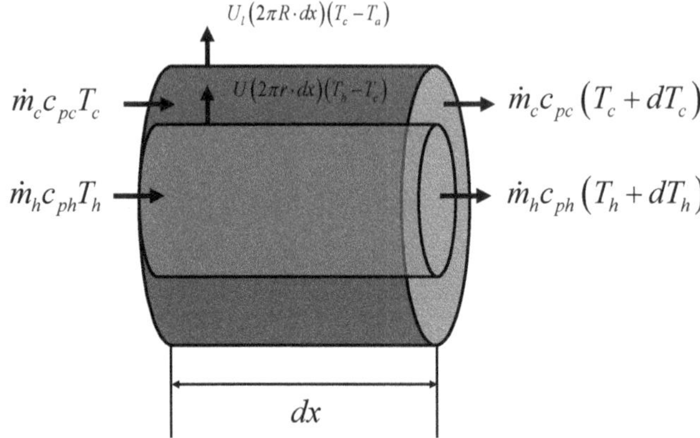

Applying law of energy conservation for differential elements of the two streams shown

$$U\left(2\pi r\cdot dx\right)\left(T_h-T_c\right)-U_l\left(2\pi R\cdot dx\right)\left(T_c-T_a\right)=\dot{m}_c c_{pc}dT_c \tag{1}$$

$$-U\left(2\pi r\cdot dx\right)\left(T_h-T_c\right)=\dot{m}_h c_{ph}dT_h \tag{2}$$

Let temperature above ambient value be denoted by θ and be called excess temperature.

Equations 1 and 2 can be expressed in terms of excess temperature as

$$\frac{d\theta_c}{dx}=a_0\left(\theta_h-\theta_c\right)-b_0\theta_c \tag{3}$$

$$\frac{d\theta_h}{dx}=-c_0\left(\theta_h-\theta_c\right) \tag{4}$$

Where $a_0 = \dfrac{2\pi r U}{\dot{m}_c c_{pc}}, b_0 = \dfrac{2\pi R U_l}{\dot{m}_c c_{pc}}, c_0 = \dfrac{2\pi r U}{\dot{m}_h c_{ph}}$. Substituting θ_c from equation 4 in 3

$$\frac{d\theta_h}{dx} + \frac{1}{c_0}\frac{d^2\theta_h}{dx^2} = -\frac{a_0}{c_0}\frac{d\theta_h}{dx} - b_0\theta_h - \frac{b_0}{c_0}\frac{d\theta_h}{dx}$$

$$\Rightarrow \frac{d^2\theta_h}{dx^2} + \left(a_0 + b_0 + c_0\right)\frac{d\theta_h}{dx} + b_0 c_0 \theta_h = 0 \tag{5}$$

Equation 5 is a second order, linear homogeneous ordinary differential equation with constant coefficients in $\theta_h(x)$. Its general solution is

$$\theta_h = T_h - T_a = C_1 e^{p_1 x} + C_2 e^{p_2 x} \tag{6}$$

Where
$$p_1, p_2 = \frac{-\left(a_0 + b_0 + c_0\right) \pm \sqrt{\left(a_0 + b_0 + c_0\right)^2 - 4b_0 c_0}}{2}.$$ Substituting Equation 6 in 4

$$\theta_c = T_c - T_a = C_1 e^{p_1 x} + C_2 e^{p_2 x} + \frac{C_1 p_1 e^{p_1 x} + C_2 p_2 e^{p_2 x}}{c_0}$$

$$\Rightarrow \theta_c = T_c - T_a = C_1\left(1 + \frac{p_1}{c_0}\right)e^{p_1 x} + C_2\left(1 + \frac{p_2}{c_0}\right)e^{p_2 x} \tag{7}$$

The unknown constants C_1 and C_2 need to be evaluated using the boundary conditions mentioned underneath

$$\left(\theta_h\right)_{x=0} = T_{hi} - T_a \tag{8}$$

$$\left(\theta_c\right)_{x=0} = T_{ci} - T_a \tag{9}$$

Feeding equations 6 and 7 in 8 and 9

$$C_1 + C_2 = T_{hi} - T_a \tag{10}$$

$$C_1\left(1 + \frac{p_1}{c_0}\right) + C_2\left(1 + \frac{p_2}{c_0}\right) = T_{ci} - T_a \tag{11}$$

Solving,

$$C_1 = \frac{\left\{c_0\left(T_{ci} - T_a\right)\right\} - \left\{\left(c_0 + p_2\right)\left(T_{hi} - T_a\right)\right\}}{p_1 - p_2} \tag{12}$$

$$C_2 = \frac{\left\{\left(p_1 + c_0\right)\left(T_{hi} - T_a\right)\right\} - \left\{c_0\left(T_{ci} - T_a\right)\right\}}{p_1 - p_2} \tag{13}$$

Feeding equations 12 and 13 in 6 and 7

$$\theta_h = T_h - T_a = \left[\frac{\left\{c_0\left(T_{ci} - T_a\right)\right\} - \left\{\left(c_0 + p_2\right)\left(T_{hi} - T_a\right)\right\}}{p_1 - p_2}\right]e^{p_1 x} +$$
$$\left[\frac{\left\{\left(p_1 + c_0\right)\left(T_{hi} - T_a\right)\right\} - \left\{c_0\left(T_{ci} - T_a\right)\right\}}{p_1 - p_2}\right]e^{p_2 x} \tag{14}$$

$$\theta_c = T_c - T_a = \left[\frac{\left\{c_0\left(T_{ci} - T_a\right)\right\} - \left\{\left(c_0 + p_2\right)\left(T_{hi} - T_a\right)\right\}}{p_1 - p_2}\right]\left(1 + \frac{p_1}{c_0}\right)e^{p_1 x} +$$
$$\left[\frac{\left\{\left(p_1 + c_0\right)\left(T_{hi} - T_a\right)\right\} - \left\{c_0\left(T_{ci} - T_a\right)\right\}}{p_1 - p_2}\right]\left(1 + \frac{p_2}{c_0}\right)e^{p_2 x} \tag{15}$$

The outlet temperature of cold fluid is value of T_c at $x=L$

$$\left(T_c\right)_{x=L} = T_a +$$

$$\left[\frac{\left\{c_0\left(T_{ci}-T_a\right)\right\}-\left\{\left(c_0+p_2\right)\left(T_{hi}-T_a\right)\right\}}{p_1-p_2}\right]\left(1+\frac{p_1}{c_0}\right)e^{p_1 L} +$$

$$\left[\frac{\left\{\left(p_1+c_0\right)\left(T_{hi}-T_a\right)\right\}-\left\{c_0\left(T_{ci}-T_a\right)\right\}}{p_1-p_2}\right]\left(1+\frac{p_2}{c_0}\right)e^{p_2 L} \tag{16}$$

Expanding back p_1, p_2

$$\left(T_c\right)_{x=L} = T_a +$$

$$\left[\frac{\left\{c_0\left(T_{ci}-T_a\right)\right\}-\left\{\left(\dfrac{c_0-a_0-b_0-\sqrt{\left(a_0+b_0+c_0\right)^2-4b_0c_0}}{2}\right)\left(T_{hi}-T_a\right)\right\}}{\sqrt{\left(a_0+b_0+c_0\right)^2-4b_0c_0}}\right]$$

$$\left(\frac{c_0-a_0-b_0+\sqrt{\left(a_0+b_0+c_0\right)^2-4b_0c_0}}{2c_0}\right)e^{\frac{-\left(a_0+b_0+c_0\right)L+L\sqrt{\left(a_0+b_0+c_0\right)^2-4b_0c_0}}{2}} +$$

$$\left[\frac{\left\{\left(\dfrac{c_0-a_0-b_0+\sqrt{\left(a_0+b_0+c_0\right)^2-4b_0c_0}}{2}\right)\left(T_{hi}-T_u\right)\right\}-\left\{c_0\left(T_{ci}-T_a\right)\right\}}{\sqrt{\left(a_0+b_0+c_0\right)^2-4b_0c_0}}\right]$$

$$\left(\frac{c_0-a_0-b_0-\sqrt{\left(a_0+b_0+c_0\right)^2-4b_0c_0}}{2c_0}\right)e^{\frac{-\left(a_0+b_0+c_0\right)L-L\sqrt{\left(a_0+b_0+c_0\right)^2-4b_0c_0}}{2}} \tag{17}$$

Expanding back a_0, b_0, c_0

$$\left(T_c\right)_{x=L} = T_a +$$

$$
\left[\frac{\left\{\dfrac{2\pi r U\left(T_{ci}-T_a\right)}{\dot{m}_h c_{ph}}\right\}-\left\{\left(\dfrac{2\pi r U}{\dot{m}_h c_{ph}}-\dfrac{2\pi r U}{\dot{m}_c c_{pc}}-\dfrac{2\pi R U_l}{\dot{m}_c c_{pc}}-\sqrt{\left(\dfrac{2\pi r U}{\dot{m}_c c_{pc}}+\dfrac{2\pi R U_l}{\dot{m}_c c_{pc}}+\dfrac{2\pi r U}{\dot{m}_h c_{ph}}\right)^2-4\dfrac{2\pi R U_l}{\dot{m}_c c_{pc}}\cdot\dfrac{2\pi r U}{\dot{m}_h c_{ph}}}\right)\left(\dfrac{T_{hi}-T_a}{2}\right)\right\}}{\sqrt{\left(\dfrac{2\pi r U}{\dot{m}_c c_{pc}}+\dfrac{2\pi R U_l}{\dot{m}_c c_{pc}}+\dfrac{2\pi r U}{\dot{m}_h c_{ph}}\right)^2-4\dfrac{2\pi R U_l}{\dot{m}_c c_{pc}}\cdot\dfrac{2\pi r U}{\dot{m}_h c_{ph}}}}\right]
$$

$$
\left(\frac{\dfrac{2\pi r U}{\dot{m}_h c_{ph}}-\dfrac{2\pi r U}{\dot{m}_c c_{pc}}-\dfrac{2\pi R U_l}{\dot{m}_c c_{pc}}-\sqrt{\left(\dfrac{2\pi r U}{\dot{m}_c c_{pc}}+\dfrac{2\pi R U_l}{\dot{m}_c c_{pc}}+\dfrac{2\pi r U}{\dot{m}_h c_{ph}}\right)^2-4\dfrac{2\pi R U_l}{\dot{m}_c c_{pc}}\cdot\dfrac{2\pi r U}{\dot{m}_h c_{ph}}}}{2\cdot\dfrac{2\pi r U}{\dot{m}_h c_{ph}}}\right)e^{\dfrac{-\left(\dfrac{2\pi r U}{\dot{m}_c c_{pc}}+\dfrac{2\pi R U_l}{\dot{m}_c c_{pc}}+\dfrac{2\pi r U}{\dot{m}_h c_{ph}}\right)L+L\sqrt{\left(\dfrac{2\pi r U}{\dot{m}_c c_{pc}}+\dfrac{2\pi R U_l}{\dot{m}_c c_{pc}}+\dfrac{2\pi r U}{\dot{m}_h c_{ph}}\right)^2-4\dfrac{2\pi R U_l}{\dot{m}_c c_{pc}}\cdot\dfrac{2\pi r U}{\dot{m}_h c_{ph}}}}{2}}+
$$

$$
\left[\frac{\left\{\left(\dfrac{2\pi r U}{\dot{m}_h c_{ph}}-\dfrac{2\pi r U}{\dot{m}_c c_{pc}}-\dfrac{2\pi R U_l}{\dot{m}_c c_{pc}}-\sqrt{\left(\dfrac{2\pi r U}{\dot{m}_c c_{pc}}+\dfrac{2\pi R U_l}{\dot{m}_c c_{pc}}+\dfrac{2\pi r U}{\dot{m}_h c_{ph}}\right)^2-4\dfrac{2\pi R U_l}{\dot{m}_c c_{pc}}\cdot\dfrac{2\pi r U}{\dot{m}_h c_{ph}}}\right)\left(\dfrac{T_{hi}-T_a}{2}\right)\right\}-\left\{\dfrac{2\pi r U\left(T_{ci}-T_a\right)}{\dot{m}_h c_{ph}}\right\}}{\sqrt{\left(\dfrac{2\pi r U}{\dot{m}_c c_{pc}}+\dfrac{2\pi R U_l}{\dot{m}_c c_{pc}}+\dfrac{2\pi r U}{\dot{m}_h c_{ph}}\right)^2-4\dfrac{2\pi R U_l}{\dot{m}_c c_{pc}}\cdot\dfrac{2\pi r U}{\dot{m}_h c_{ph}}}}\right]
$$

$$
\left(\frac{\dfrac{2\pi r U}{\dot{m}_h c_{ph}}-\dfrac{2\pi r U}{\dot{m}_c c_{pc}}-\dfrac{2\pi R U_l}{\dot{m}_c c_{pc}}-\sqrt{\left(\dfrac{2\pi r U}{\dot{m}_c c_{pc}}+\dfrac{2\pi R U_l}{\dot{m}_c c_{pc}}+\dfrac{2\pi r U}{\dot{m}_h c_{ph}}\right)^2-4\dfrac{2\pi R U_l}{\dot{m}_c c_{pc}}\cdot\dfrac{2\pi r U}{\dot{m}_h c_{ph}}}}{2\cdot\dfrac{2\pi r U}{\dot{m}_h c_{ph}}}\right)e^{\dfrac{-\left(\dfrac{2\pi r U}{\dot{m}_c c_{pc}}+\dfrac{2\pi R U_l}{\dot{m}_c c_{pc}}+\dfrac{2\pi r U}{\dot{m}_h c_{ph}}\right)L-L\sqrt{\left(\dfrac{2\pi r U}{\dot{m}_c c_{pc}}+\dfrac{2\pi R U_l}{\dot{m}_c c_{pc}}+\dfrac{2\pi r U}{\dot{m}_h c_{ph}}\right)^2-4\dfrac{2\pi R U_l}{\dot{m}_c c_{pc}}\cdot\dfrac{2\pi r U}{\dot{m}_h c_{ph}}}}{2}}
$$

$$\tag{18}$$